International Conference "Actual Problem of Electromechanics and Electrotechnology" (APEET-2017)

IOP Conference Series: Materials Science and Engineering
Volume 313

Yekaterinburg, Russia
13 – 16 November 2017

ISBN: 978-1-5108-5890-9
ISSN: 1757-8981

Printed from e-media with permission by:

Curran Associates, Inc.
57 Morehouse Lane
Red Hook, NY 12571

Some format issues inherent in the e-media version may also appear in this print version.

Printed by Curran Associates, Inc. (2018)

For permission requests, please contact the Institute of Physics
at the address below.

Institute of Physics
Dirac House, Temple Back
Bristol BS1 6BE UK

Phone: 44 1 17 929 7481
Fax: 44 1 17 920 0979

techtracking@iop.org

Additional copies of this publication are available from:

Curran Associates, Inc.
57 Morehouse Lane
Red Hook, NY 12571 USA
Phone: 845-758-0400
Fax: 845-758-2633
Email: curran@proceedings.com
Web: www.proceedings.com

Table of contents

Volume 313

International Conference "Actual Problem of Electromechanics and Electrotechnology"

13–16 November 2017, Yekaterinburg, Russian Federation

Accepted papers received: 1 February 2018
Published online: 19 February 2018

Preface

Papers

IOP Conf. Series: Materials Science and Engineering **313** (2018) 011001 doi:10.1088/1757-899X/313/1/011001

1. Preface

International conference "Actual Problems of Electromechanics and Electrotechnology APEET-2017" was held November 13-16, 2017 in the Ural Federal University named after the first President of Russia B.N.Yeltsin.

This conference presents papers that largely reflect problems that scientists and specialists in the field of electromechanics and modern electrotechnology solve, as well as in related fields of technology. This applies, first of all, to electrothermal equipment and electrotechnologies, power supplies and power supply systems of this equipment.

Other part of the reports is devoted to the development and investigation of electromechanical energy converters, including permanent magnetic motors, alternating current motors, linear and stepper motors, special electric machines. Much attention is paid to energy-saving aspects in electrotechnical and electromechanical systems, as well as training specialists.

Conference was held with the support of the Department «Electrical engineering and electrotechnology systems» and the Department "Electrical Machines" of the Ural Power Engineering Institute.

2. Lists of committees/participants

Organizing Committee

Chairmen

Sarapulov F.N., Professor, Ural Power Engineering Institute, Ural Federal University

Co-Chairmans

Plastun A.T., Professor, Ural Power Engineering Institute, Ural Federal University
Frizen V.E., Ural Power Engineering Institute, Ural Federal University
Sarapulov S.F., Ural Power Engineering Institute, Ural Federal University

Members of Committee

Denisenko V.I., Professor, Ural Power Engineering Institute, Ural Federal University
Konyaev A.U., Professor, Ural Power Engineering Institute, Ural Federal University
Bytchkov S.A., Ural Power Engineering Institute, Ural Federal University
Tarasov F.E., Ural Power Engineering Institute, Ural Federal University
Bolotin K.E., Ural Power Engineering Institute, Ural Federal University
Shvydky E.L., Ural Power Engineering Institute, Ural Federal University
Smolyanov I.A., Ural Power Engineering Institute, Ural Federal University
Abdulaev Zh.O., Ural Power Engineering Institute, Ural Federal University
Antipin A.S., Ural Power Engineering Institute, Ural Federal University
Rodinov I.E., Ural Power Engineering Institute, Ural Federal University

IOP Conf. Series: Materials Science and Engineering **313** (2018) 011002 doi:10.1088/1757-899X/313/1/011002

Peer review statement

All papers published in this volume of *IOP Conference Series: Materials Science and Engineering* have been peer reviewed through processes administered by the proceedings Editors. Reviews were conducted by expert referees to the professional and scientific standards expected of a proceedings journal published by IOP Publishing.

IOP Conf. Series: Materials Science and Engineering **313** (2018) 012001 doi:10.1088/1757-899X/313/1/012001

The efficiency of the use of composite materials in electrotechnical equipment

K Kim[1] and S Ivanov[2]

[1]Professor, Emperor Alexander I St. Petersburg State Transport University, Saint Petersburg, RU
[2]Professor, Komsomolsk-na-Amure State University, Komsomolsk-na-Amure, RU

E-mail kimkk@inbox.ru

Abstract. The indicators of the efficiency of electrical installations are directly connected with the creating and using of new composite materials with the desired performance properties. The practical application of composite materials is one of the perspective scientific and technical directions, providing the increase of the efficiency of electrical installations due to the sealing of current parts by protecting them from the external medium. The technical characteristics of the composite material match to its structure and depend on the properties of the individual components. The verification of the compliance of material parameters is implemented by the methods of the computer analysis of a model of composite material in the form of the structure in which the individual elements have thermodynamic properties of the corresponding phase state. In the study the topology of individual elements in the material structure is defined by the conditional boundaries of the section within the studied composite. The efficiency of using the composite materials includes the raising of electrical safety, increasing the durability, reducing the costs of maintenance and repair and the extension of the scope of installations.

1. Introduction

The economic analysis of the efficiency of the Russian energy facilities shows that the energy per unit of the manufactured products is consumed on average by 2...4 times more than in the other industrialized countries. The indicators of Russian energy efficiency are lower than China's indicators 1.8 times, USA 2 times, Japan 6 times [1]. It is important to note that the possibility of a complex energy saving is primarily connected with minimizing the level of costs for its realization. The main reason of the current state of the low efficiency of electrical installations is the deterioration of the main power equipment, its level lags behind the foreign equipment by 30...40 years.

According to the expert estimations the costs for the implementation of energy saving technologies is almost by 4 times lower than the costs of production, distribution and transportation of the produced energy to the places of consumption. That is, in practice, in the difficult economic conditions there is a problem of increasing of energy saving and energy efficiency, but not increasing new powers which are ineffective due to the low level of the used technologies and outdated equipment.

2. Statement of the research problem

The analysis of the strategy of implementing energy efficiency programs shows that firstly we should implement technical measures not requiring a great investment. To such measures we relate the modernization of equipment, which has a maximum energy losses, for example, the complexes for transporting thermal energy their losses often exceed 50 %. This is confirmed by the comparison of different energy saving technologies, showing that the main attention should be focused on the problems of using all the types of energy with the minimal energy losses due to, firstly, the increasing of the efficiency and energy excellence of equipment for the generation and transportation of thermal energy.

The reduction of losses in such equipment provides both only improving the operating characteristics (efficiency, power factor, dimensions and weight) and also the characteristics which are determined by the interaction of the technical objects and external environment such as the technosphere safety, system efficiency and compatibility. The increasing of the equipment energy efficiency is possible due to the improving individual elements, process or system modifications and choice of software and hardware management tools. We achieve a substantial result by improving the reliability of electrical devices too.

One of the perspective directions of increasing efficiency of energy devices is using the hermetic designs. The results of investigating the special hermetic electromechanical converters which we have nowadays, confirm the prospect of their using [2]. A typical example of such development is the hermetic AC electric drive of the compressor of 20 MW produced by Siemens to transport the associated gases from the great depths [3]. The device effectiveness is determined by the operating time not less than 20 thousand hours. It should be noted that in the heavy-load conditions the hermetic device is almost the only really workable variant of a drive device. Its feasibility is provided by the high reliability of the design. The scope of using the hermetic AC drives includes, as the already mentioned compressors and also the technological centrifuges, mixers, pumps and apparatus which should guarantee the total zero leakage of reactive and polluting components into the environment, the operation with flammable mediums, the operation at the high and low pressures and temperatures.

The increasing of the energy efficiency of technical systems using a hermetic equipment is also determined by the possibility of increasing electromagnetic loads due to the transition to the direct cooling of the elements.

The practical realization of such electrical installations is connected with the development and production of composite materials with the desired physical and chemical characteristics that is with one of perspective scientific and technical directions, affecting the strategic interests of all the industrialized countries. The improving efficiency of the electrical installation is provided by the hermetic sealing of the current parts due to their capsulation. The capsulation is that, for example, the insulation of the stator winding is protected from the external medium by a molded shell made of a composite material. This allows to provide a higher safety class of the installation, to increase the resource, to simplify the design, to increase its structural reliability, to reduce the cost of maintenance and repair, to extend the range of operating modes.

The technological basis of capsulation is the implement of special processes, providing the creation of composite materials with the desired performance properties. In this case the technological process significantly affects the physico-chemical characteristics of the composites dependent on the quantitative and qualitative composition of the elements forming the structure.

The integral characteristics of the composite material are determined by its structure and the properties of some components. In the absence of the exact mathematical model of the researched object, which is a composite material, the providing of the desired properties can be achieved with simulation methods. The simulation model can be built on the basis of the physical understanding about the researched object. The physical model of composite material is reduced to the highly dispersed structure in which some components are in the state of aggregation, at that the colloid particles have the thermodynamic properties of the corresponding phase state. The capsulated material is heterogeneous after the polymerization process. It is characterized by the relative insolubility of the components having the determined interfaces within the researched object. The definition of topology of individual elements in the material structure is connected with the solution of the Van der Waals equation, which allows to take into account the factors of influence of intermolecular action of the non-chemical origin.

It should be noted that the cost of the hermetic electrical installations is more than the cost of the similar standard equipment by 30 % however, this is not a determining factor in some cases because of the economic effect achieved by improving the reliability, security and durability of the installation.

The capsulation of installations by the composite materials significantly increases the maintenance intervals at the simultaneous increasing of the mean-time-between-failures. According to the statistics of the company HERMETIC-Pumpen GmbH the average mean-time-between-failures of the circulation pumps with screening motors is 7.5 years, with the traditional ones average time is 2 years.

The electrical installations using insulating composite materials are characterized by the higher electrical safety. The hermetical implementation of the output boxes, wire cable and other elements eliminates the access of gas and liquid and also prevents the emission of harmful substances into the atmosphere. It allows to use them in the food and medical industries. The capsulation allows to provide the required security when operating in the aggressive or toxic mediums. The using of composite materials increases the resistance of the installation to the mechanical effects and leads to the lower of the levels of self vibrations and noise. In addition, the hermetic installations are more compact that allows to place them in the hard-to-reach places. Thus the task of evaluating the effectiveness of using composite materials in the combined heat generating plants is the consideration of their design features at designing and developing recommendations for the practical implementation of this class of devices.

3. An example of the implementation

We consider the use of the composite material in the electromechanical converter of the alternating current (the asynchronous motor with a modified squirrel-cage rotor). The actuality of such a technical solution is connected with the possibility of using the frequency control allowing not only effectively to regulate the speed of rotation but also to provide the energy saving modes of operation due to the optimized algorithms of the electricity consumption using intelligent technologies (fuzzy logic, neural networks, neuro approaches).

The use of the composite as the material capsulating the stator windings allows to realize the modes of the generating and transporting thermal energy in the same device, it is the basis of creating a new class of devices – the combined heat generating converters [4-9].

One possible design of the stator of an electrical plant using two types of composite materials is shown in Figure1, where 1 is the antifriction material which provides the sliding of the hollow rotor and 2 is the material which insulates the stator winding.

Figure1. The example of using composite materials: 1 is the self-lubricating antifriction material; 2 is the insulating material.

Despite the structural similarity to the usual electric machines of the alternating current, these heat generating devices with the composite coating of the stator have a number of fundamental differences such as the presence of a motionless heating element fixed on the capsulated stator and a rotatable hollow non-magnetic rotor. The pressure blades are fixed inside this rotor. The necessity of using a motionless heating element connected with the fact that in the modes of the device close to the synchronous modes the amount of heat losses generated in the rotor is greatly reduced due to the convergence of the speeds of magnetic field rotation and the rotor. Therefore, to ensure the required heat output there is an additional source in the design of the device. The power dissipation of this source is slightly depends on the speed of the rotor [10].

The use of composite material allows to provide a low flow resistance due to the lack of a traditional shaft and bearings. It should be noted that in the absence of ferromagnetic elements in the inner bore of the stator, the heat generating device is characterized by a significant magnetizing current and the reactive power consumption causing significant heating of the composite material. The value of the heat load of the material depends on the dimensional relationships between the magnetic core and the air gap. Their values determine the hydraulic resistance of the flow part of the device. Since the value of the main magnetic flux determines the magnetic induction and the minimum width of the stator tooth, the thickness of the inner magnetic core is limited from the bottom. In particular, for the studied variant of the device with power 2.2 kW the width of the stator tooth is approximately equal to 4 mm but the reluctance is significantly reduced even when the thickness of element is 2 mm.

The axial length of the inner magnetic core is determined more difficult as it both changes the distribution of magnetic field in the inner area of the device and changes the process of heat transferring of an additional heat source due to the change of the hydraulic resistance. The results of the research showed that we can determine the relative range, in which we achieve the decreasing of the resistance of the magnetizing circuit without the significant deterioration of the heat transfer. This range is approximately the 0.25...0.70 of the air gap length.

In the analysis of electromagnetic processes determining the operating characteristics of the composite material the device can be represented in the form of a rotating AC converter with two secondary windings: motionless and moving hollow shells. For the description of electromagnetic processes at the constant distributions of magnetic field and power flows the real converter having a number of the turns of the primary winding w_1 and the secondary windings with w_2 and w_3 is replaced by the equivalent converter with the number of turns of the secondary windings equal to the number of turns of the primary winding. The impedances of the secondary windings are changed for the invariance of the active and reactive power components.

For the specified voltage the currents in all the electrical circuits are theoretically determined by their impedances. The primary winding includes the active r_1 and inductive x_{1L} resistances, the motionless secondary winding includes the reduced active r'_2, inductive x_{2L} and capacitive x_{2C} resistances, the rotating secondary winding includes the reduced active r_{3S}, inductive x_{3L} and capacitive x_{3C} resistances. However, in the real converters, the thickness of the walls of the hollow non-magnetic rotor is many times less than the depth of penetration of the electromagnetic field, therefore, we can assume that the inductive resistances of leakage of the hollow rotors x'_{2L}, x'_{3L} are small and the eddy currents are distributed throughout the thickness of the rotor walls almost uniformly at all the slides, so we can assume that the resistance r'_{3S} doesn't depend on sliding. This allows us to find approximately a load component of the current consumption and the electromagnetic power transmitted from the stator and providing two processes – the heat of the heat carrier by the motionless element and the heat and transportation of the heat carrier by the rotating rotor [11].

The mechanical power of the device is determined by the power density perceived by the surface of the rotor as the product of the tangential stress on the tangential circumferential speed. It provides the pressure and performance. The heat power equal to the sum of all the heats provides the temperature of the heat carrier. It is obvious that it is necessary to define electric, magnetic and additional losses in the converter to define the thermal power.

The synthesis of the device and the frequency converter requires to take into account the nonuniform distribution of the current density over the cross section of the conductors of the primary winding due to the skin effect. Therefore, the determination of the electrical losses with the help active resistance obtained for the uniform distribution of the current density over the cross section that is only true for direct currents and currents of low frequency (several Hertz).

The nonuniform distribution of current depends on the value of the leakage field in the groove, its field lines are almost perpendicular to the axis of the groove, thereby the coupling of the field with the conductor elements located at the same height of the groove, defines the similar inductive resistances of such elements. At the same time, the inductive resistances of the elements located at the different height in the groove, will be different. The increase of the current density occurs as approaching to the surface of the groove, accordingly, the useful section is reduced and the active resistance increases. In practice for the quantitative evaluation we use the magnification ratio of the winding resistance due to the skin effect, which is dependent on the frequency of current and the geometry of the conductor.

The define of magnetic losses requires the consideration of technological factors, a type of magnetization reversal (pulsation or rotating of the field) and additional components. In practice we use the specific magnetic losses for the calculation, for example, the measured losses experimentally at the frequency of 50 Hz and magnetic induction of 1 Tesla. The recalculation to other frequencies and inductions requires the separation of hysteresis losses, proportional to the frequency, and the eddy-current losses currents depend on the square of frequency of the supply voltage.

The increase of the magnetic losses due to the reasons of the technological nature is taken into account by the empirical coefficients of the technological losses for the characteristic parts of the magnetic circuit. The total ratio of the technological losses for the yoke of the magnetic circuit is 1.3...1.6 and for teeth it is 1.7...1.8. The additional magnetic losses due to the non-sinusoidal from the higher harmonics are taken into account by the ratio of the added losses (1.2...1.3).

When we conducted the research we used the automated information-measuring complex based on a set of Labview virtual instruments, the system of T-FLEX CAD parametric design, the statistical software complex Statistica, CFD SolidWorks Flow Simulation, the software developed in the environment of Delphi and Elcut.

The main source of heating of the heat carrier is set by the value of electric losses in the additional heating element.The analytical study of electromagnetic fields taking into account their shape, the laws of the current distribution and the physical properties of the mediums suggests a number of assumptions. At the analysis of fields we consider only the first space harmonics of the magnetizing force; the value of the air gap is many times smaller than the external diameter; the current density is constant through the thickness of the rotor; the distribution of field intensity has only a normal component; the magnetic permeability of steel is infinite; the induction is outside the stator package is equal to zero. To summarize the fields in the current and current-free areas we introduce a vector potential. Let’s write Maxwell's equations, then Laplace's equation for current-free areas, and Poisson’s equation for the area of the eddy currents. To solve these equations we determine the vector potential, the field and density of the eddy current. The electric power produced in the heating element is determined by the real part of Poynting’s vector.

The definition of the additional losses is connected with the necessity of considering the influence of the higher harmonics of electromagnetic field and currents in the secondary circuits. The additional losses are taken place both in a motionless magnetic core and in low magnetic elements (the secondary windings, end caps, etc). Practically, these losses are estimated approximately due to their relatively small values, but mostly it is due to the high complexity of their exact calculation. The losses arising from the movement of the heated medium are divided into hydraulic and mechanical. The accurate determination of these losses is necessary to find the temperature of the composite material. At the engineering calculations these losses are accounted by the coefficients depending on the structural parameters and operational factors similar to the calculation of the pressure characteristics of fans, compressors or pumps. To check the obtained results and recommendations we use the package Comsol Multiphysics which is an interactive environment for modeling and calculation of the task reduced to the partial differential equations by the finite element method.

In calculating the magnetic field we use a two-dimensional model of the magnetic field of the alternating current. The problem is reduced to a differential equation in private derivatives relatively to the complex value of the z-component of the magnetic vector potential. This analysis confirms the possibility of using composite materials for electrical devices of heating and transport of the heat agent providing a fairly stable resulting temperature of the heated medium at the rated supply voltages, frequency and the given heat-transfer coefficient in the range of slips 0.4...0.9, which corresponds to the operation mode of the converter with the high resisting moment arising while the movement of the viscous working medium. As the estimation of effectiveness of using the composite materials in the research plant we selected the growth rate of the defect formation of the turn winding insulation of the capsulated stator depending on the constructive and operational parameters. The dependence of the growth rate of the defect formation (G_r) on the structural parameters (C_f is the coefficient of filling of the groove, l_{end} is the length of the end winding) is shown in Figure 2.

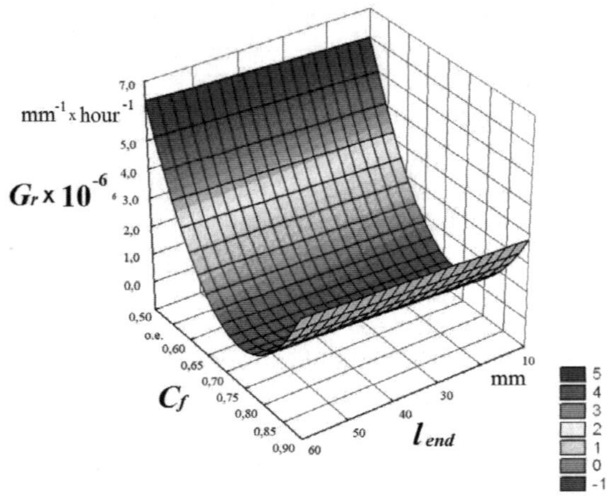

Figure 2. The dependence of the growth rate of the defect formation on the structural parameters

The comparative analysis of capsulated and traditional designs showed that the rate of defect formation in the designs made from the composite materials is less. Let's list the equity contribution of separate factors influencing on the plant reliability: the influence of the diameter of the winding wire is 3 %; the influence of the value of the end winding is 4 %; the vibration acceleration is 2 %; the influence of the fill factor of the groove is 6 %; the influence of the insulation quality of the is 10 %; the influence of the winding temperature is 75 %.

4. Conclusion

The efficiency of using the electrical equipment is directly determined both by the value of the losses in this equipment and also by their level of reliability. The problems of reliability are important especially in the autonomous complexes, therefore with improving the production, increasing requirements to the equipment reliability must be obligatory. The comparative analysis showed the obvious advantages of using the composite materials to ensure the requirements of safety and reliability. The use of electrical installations with the use of new composite materials allows to reduce the costs of repair and restoration and in general it can be considered as one of the measures to increase their efficiency when we modernize industrial and agricultural companies.

References

[1] Bogdanov A.B. 2010 *Energy-saving* **5** p.46-53.

[2] Kim K.K. and Ivanov S.N. 2011 *Electromechanical generators of thermal energy* (LAMBERT Academic Publishing).

[3] Durymanov V.V.et all 2010 *Turbines and diesel engines* **2** p. 10-15.

[4] Ivanov S.N. and Skripilev A.A. 2016 *Proceedings of the KnAGTU* **III-1(27)** p. 20-26.

[5] Patent RF, no 137163, 2014.
[6] Patent RF, no 150936, 2015.
[7] Patent RF, no 154737, 2015.
[8] Patent RF, no 2410852, 2011.
[9] Patent RF, no 2525234, 2014.
[10] Ivanov S.N. 2016 *Proc. XX International scientific and technical conference* (*Tula*) p. 80-81.
[11] Kim K.K. 2008 *Electrical and magnetic circuits* (Oxford, UK Cooxmoor Publishing Company).

Non-conventional rule of making a periodically varying different-pole magnetic field in low-power alternating current electrical machines with using ring coils in multiphase armature winding

A T Plastun [1], O V Tikhonova [2], I V Malygin [3]
[1] Professor, "Electrical machines" department, Ural Federal University, Yekaterinburg, Russia
[2] Graduate student, "Electrical machines" department, Ural Federal University, Yekaterinburg, Russia
[3] Assistant, "Electrical machines" department, Ural Federal University, Yekaterinburg, Russia

E-mail: a.t.plasun@urfu.ru

Abstract. The paper presents methods of making a periodically varying different-pole magnetic field in low-power electrical machines. Authors consider classical designs of electrical machines and machines with ring windings in armature, structural features and calculated parameters of magnetic circuit for these machines.

1. Introduction

As it is known there are two ways for making of periodically varying different-pole magnetic field in low-power alternating current machines [1].

First of them is connected with case when environment for formation of the magnetic field is homogeneous in magnetic position and the specific magnetic permeability at each point of the considered region is, for example, one.

In this case, the formation of a periodically changing different-pole magnetic field is carried out by means of conductors with a periodic alternation of current directions in adjacent conductors, which are placed, for example, on an imaginary cylindrical surface, that is fixed in space. Conductors with counter-directed movement of electrons are connected in turns, coils, coil groups, forming a phase.

In the general case, the turns with current can have any shape (boat, rectangle, trapezoid, circle, etc). Current loops can be located on a closed or unclosed surface stretched over an imaginary cylinder (sphere, cone, screw, etc.)

When an electric current flows through the phase, different-pole induction changes of the magnetic field of a given direction with given amplitude at each point of the surface are formed. This principle is used in synchronous machines of general industrial execution with the purpose of creating an excitation field.

Conductors with a current can be placed on a plane, for example, along an arbitrarily curved line. Conductors with counter-directed movement of electrons are connected in turns, coils, coil groups, forming a phase. When an electric current flows through the phase, an alternating change in the magnetic field is formed. This method of placing phase coils is used to create an excitation field in synchronous machines, for example, linear performance.

If conductors with current are placed in a plane but along an imaginary circle, and conductors with counter-directed movement of electrons are connected into turns, coils, coil groups, forming a phase, then when the current flows through the phase an alternating different-pole change of the end-face type magnetic field is formed. This method of placing the phase coils is used to create an excitation field in synchronous machines of the end-face performance.

If one ring coil is placed on an imaginary cylindrical surface so that the geometrical axis of the coil coincides with the rotation axis of an imaginary cylindrical surface and the projection of the current density vector in each element of the coil conductor to the rotation axis of an imaginary cylindrical surface is zero, unipolar magnetic field is generated relative to imaginary surface.

If several coils of the ring type are placed on the imaginary cylindrical surface with alternating current directions in the coils, then it is known that a magnetic field is generated. The field is unipolar with respect to the imaginary surface and alternating with respect to the axis of rotation.

The second principle of obtaining periodically varying different-pole magnetic field relative to imaginary cylindrical surface refers to the case when the environment for formation of the magnetic field is non-uniform and the specific magnetic permeability at each point of the region is determined. In the region under consideration, volumes are identified where the specific magnetic permeability is one, for example air, and volumes where the specific magnetic permeability is greater than one (magnetic masses), for example, electrical steel.

There are known designs of magnetic systems of low-power electric machines in which alternating variation of the different-pole magnetic field of a given direction at each point with given amplitude is obtained with the help of a single ring-type coil and magnetic masses covering the current coil [1]. The magnetic system is claw-shaped. The geometric axis of the magnetic system and the geometrical axis of the coil coincide with the axis of an imaginary cylindrical surface, relative to which a different-pole magnetic field is created.

2. The traditional principles of the formation of a periodically varying different-pole magnetic field moving with respect to an imaginary cylindrical surface

According to the first principle of forming a periodically varying different-pole magnetic field in alternating current electric machines, for making of different-pole magnetic field moving around the rotation axis of imaginary cylindrical surface (m-1) additional phase must be placed on the same surface (m-number of phases).

The constructions of the turns, coils, coil groups and connections of neighboring phases are geometrically symmetrical and identical. The geometric axis of each coil of the phase coincides with the axis of symmetry of the magnetic field created by this coil and is perpendicular to the rotation axis of an imaginary cylindrical surface.

The geometric axes of the first coils of adjacent phases are shifted in space relative to each other by an angle of $360/m$ electric degrees. The currents in the additional phases are shifted relative to each other also by $360/m$ electric degrees by a current source, for example, an m-phase synchronous generator.

This way of forming a periodically varying different-pole magnetic field moving with respect to an imaginary cylindrical surface is used for creating a rotating magnetic field, for example, in induction motors of general industrial design.

The minimum possible number of coils in this windings is pm, where p is the number of pairs of poles. This winding is called a winding of the drum type.

A drum winding deployed on a plane along an imaginary line is used to create a linearly moving excitation field in linear AC machines.

For the organization of different-pole magnetic field moving along an imaginary cylindrical surface, obtained by placing a single coil of the ring type and alternating the geometry of the magnetic mass of a claw-shape form, the movement of the magnetic field relative to the imaginary cylinder can be created in a conventional method, for example, by rotation. The method for creating a different pole magnetic field is used, for example, in low power synchronous generators to create an excitation field moving relative to the stator, with the goal of creating an electromotive force in alternating current windings.

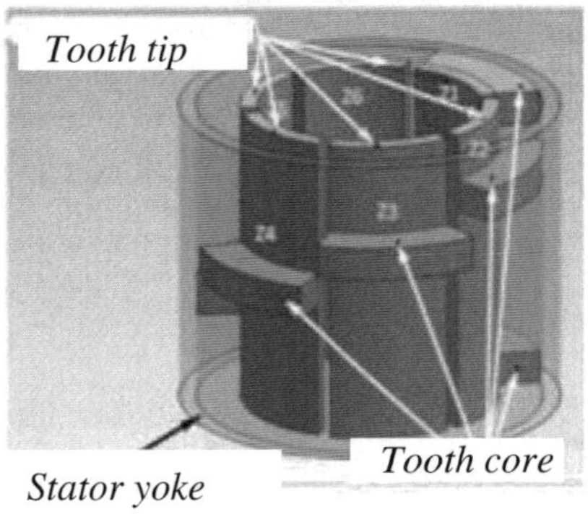

Figure 1. Geometric forms of magnetic masses in space and the order of their conjugation for p=1

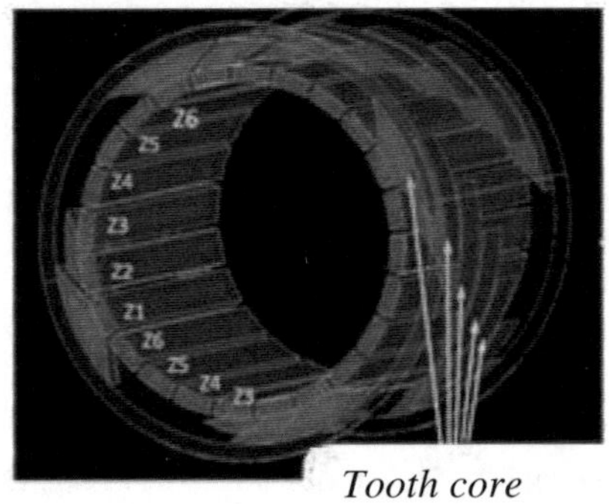

Figure 2. Geometric forms of magnetic masses in space and the order of their conjugation for $p = 4$

3. The non-traditional principles of the formation of moving magnetic field in armature of low-power electrical motors with using ring coils in multiphase armature winding

For low-power electric motors having "strewed" windings, the principle of creating of the different-pole magnetic field moving around rotation axis of imaginary cylindrical surface is observed in [2]. This way is connected with placing $2m$ ring coils and with alternating of the magnetic masses geometry of different configurations. The geometric axes of all the coils of the ring type of the armature multiphase winding according to [2] coincide with the rotation axis of the imaginary surface, relative to which a moving different-pole magnetic field is created. Magnetic masses cover each of the coils and have a different configuration.

The rule of the magnetic masses placement in space relative to current conductors in ring coils of armature winding and the order of magnetic masses conjugation are directed to a step-by-step transformation of the type of displacement of the magnetic field of the armature with respect to an imaginary cylindrical surface (rotor), namely, the spiral movement of the magnetic field in the stator yoke into radial screw motion in the zone of the tooth cores and in circular motion with respect to the imaginary surface in the zone of the tooth tips.

Figures 1-2 show the geometric forms of magnetic masses, the placement in space and the order of their conjugation for the case when the number of pole pairs is $p = 1$.

Geometric forms of magnetic masses have the following types: hollow cylinder (stator yoke), curved prism (tooth core), and segment of hollow cylinder (tooth tip).

According to figure 1, the second tooth core is shifted relative to the first tooth core by 60 electric degrees and by one sixth of the axial length of the stator. The third tooth core is shifted relative to the second tooth core by 60 electric degrees and by one sixth of the axial length of the stator, and so on. The upper (first) and lower (seventh) tooth cores are shifted relative to each other by 360 degrees.

The first tooth tip, the first tooth core, the stator yoke and the seventh tooth core are conjugated and form a short-circuited magnetic ring. The second tooth tip is shifted relative to the first tip by 60 electrical degrees. The internal surfaces of the tooth tips are conjugated with an imaginary cylindrical surface, relative to which a moving periodically varying different-pole magnetic field is formed. The inner surfaces of the tooth tips are equal.

Figure 3 shows the placement of phase ring coils with respect to magnetic masses. The connections scheme of ring type coils is shown in figure 4.

For this case and for any number of poles pairs, the minimum number of coils of a three-phase armature with ring coils is six when constructing a sixty degree phase winding zone.
It is should be noted that in the stator of an classical low-power electric motor with a single-layer three-phase armature winding, the minimum number of coils placed in the stator slots is a function of the number of pole pairs: f (p) = 3*p*.

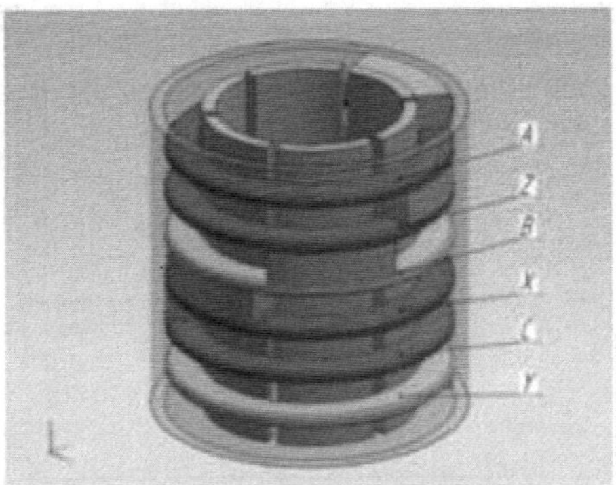

Figure 3. Placement of phase ring coils with respect to magnetic masses

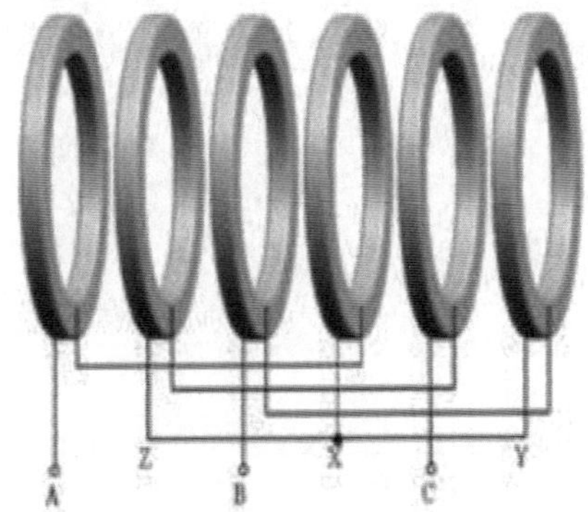

Figure 4. Connections scheme of ring type coils

4. Distribution of magnetic flux in the magnetic system of induction motor with ring windings during flow of three-phase sinusoidal currents system along armature winding

The field lines of mutual induction of a two-pole induction motor of general industrial design with the number of poles $2p = 2$ and the number of slots per pole and phase $q = 1$ are presented in simplified form in figure 5.

The magnetic field is created by the phase currents of the three-phase armature winding of the stator of an induction motor of general industrial design for the case when the current in phase A is equal to the current in phase B and equal to ½ of the current of phase C.

The field lines of mutual induction of a two-pole induction motor of general industrial design with the number of poles $2p = 2$ and the number of slots per pole and phase q = 6, when the current in phase A is equal to the current in phase B and equal to ½ of the current of phase C are presented in simplified form in figure 6.

Magnetic field vectors of a two-pole induction motor with ring windings with the number of poles $2p = 2$ and the number of slots per pole and phase q = 1 and when the current in phase A is equal to the current in phase B and equal to ½ of the current of phase C are presented in simplified form in figure 7.

IOP Conf. Series: Materials Science and Engineering **313** (2018) 012002 doi:10.1088/1757-899X/313/1/012002

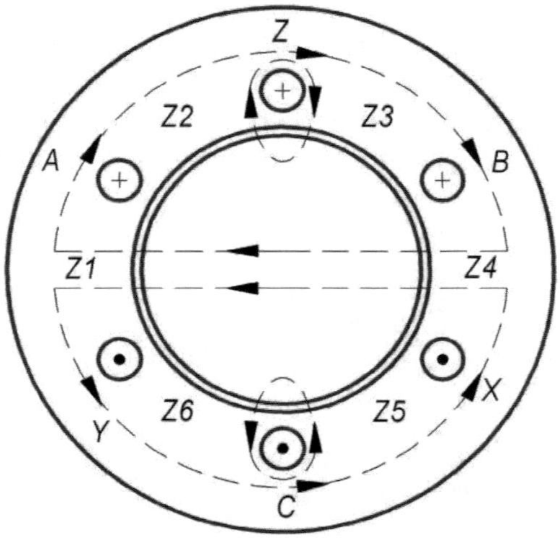

Figure 5. Field lines of mutual induction for $2p$=2, q=1

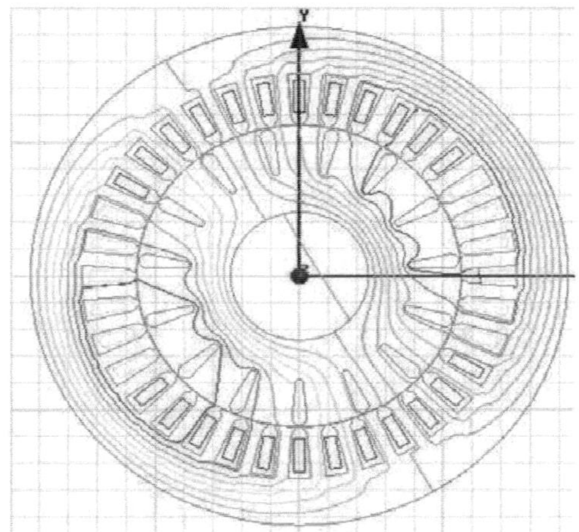

Figure 6. Field lines of mutual induction for $2p$=2, q=6

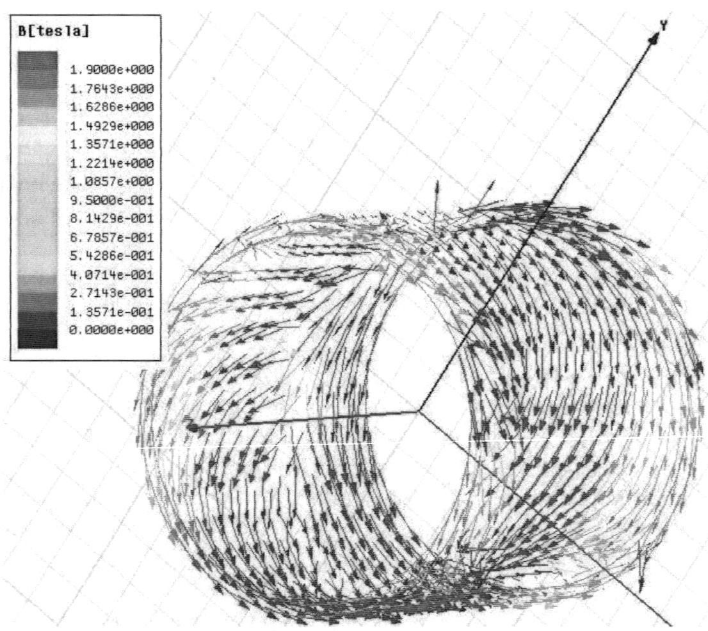

Figure 7. Vectors of magnetic field for $2p$=2, q=1 and for ring windings of armature

As it is known, the magnetic field in the magnetic system of stator and rotor of the classical design induction motor is axis-symmetric (figures 5-6).

You can see from figure 7 the magnetic field in the stator of an induction motor with ring windings in the stator yoke zone and the tooth cores has a helical shape along the axis of rotation. The helical character of the magnetic field distribution is given by the helical character of the magnetic masses distribution (see figure 1, figure 4).

Using "ELCUT" program, we determine the distribution of the flow in the working air gap of the induction motor with ring windings. The AO2-32-6 motor was chosen as the prototype design. In the calculation model of the induction motor with ring windings, the grade of electrical steel, the number of turns in the phase of the stator armature winding, the geometric dimensions of the rotor, the size of the mechanical gap were assumed to be equal to the corresponding values of the prototype. Unlike the prototype, for which the number of slots per pole and phase is two, in the design model of the induction motor with ring windings, the number of slots per pole and phase q in the first stage of calculating the flux induction distribution in the air gap along the rotor was assumed as q = 1. The current in phase A is assumed to be the no-load current of the prototype. The design of the magnetic system is collapsible. The additional air gap δΔ between the tooth core and the tooth tip is assumed equal to the value of the mechanical air gap: δΔ = 0.3 mm. The areas of all six tooth tips on the side of the air gap are equal to each other.

We define the planes that cut the magnetic system along the axis of rotation as shown in figures 8 a, b, c. For each section we construct a calculation model using the capabilities of "ELCUT" (figure 9.) And then we will perform and present the calculations.

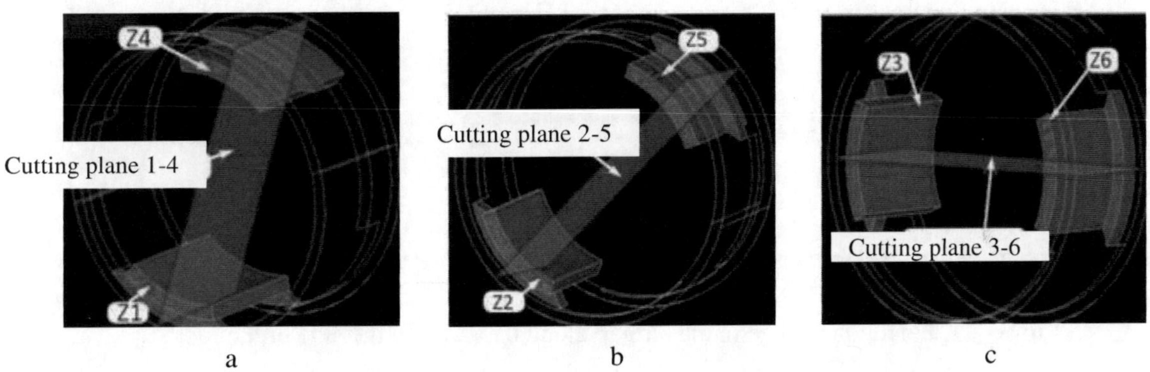

Figure 8: a, b, c. Magnetic system cutting by special planes

We construct the graphs of the induction change along the cutting planes in the air gap. Let us consider a plane passing through the tooth tips Z1 and Z4 (figures 8a). Figure 9 shows the calculation model corresponding to the longitudinal section of the magnetic circuit shown in figure 8a. Graphs of the induction change in the air gap along the rotor on the side of the tooth tips (respectively, Z1 - Z4) are shown by figures 11-12.

According to figures 9-10 you can see resulting direction of the magnetic flux lines in the air gap from the tooth tip № 4 to the tooth tip № 1 for given values and directions of the phase currents and placement of the phase coils in the magnetic system along the axis rotation, indicated in figure 9. It is presented for simplified image of magnetic circuit in cross-section.

As you can see from figures 11-12, the average value of the induction Bav.z1 in the gap under the tooth tip Z1 is equal to Bδav.z1 = 0.237 T. The average value of induction Bδav.z4 in the gap under the tooth tip Z4 is equal to Bδav.z4 = 0.247 T, which is less than 5% more than the average value of induction Bδav.z1 in the gap under the tooth tip Z1. The inequality of the average values of the

IOP Conf. Series: Materials Science and Engineering **313** (2018) 012002 doi:10.1088/1757-899X/313/1/012002

inductances in the gap under the tooth tips Z1 and Z4 is due to the assumptions made when replacing the three-dimensional model of a two-dimensional magnetic system.

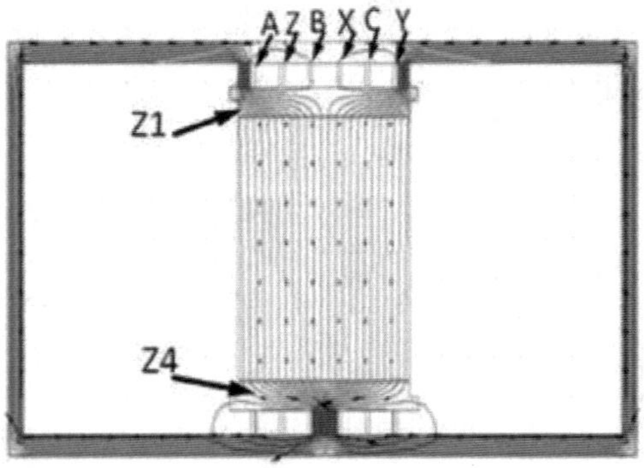

Figure 9. Calculation model corresponding to the longitudinal section of the magnetic circuit

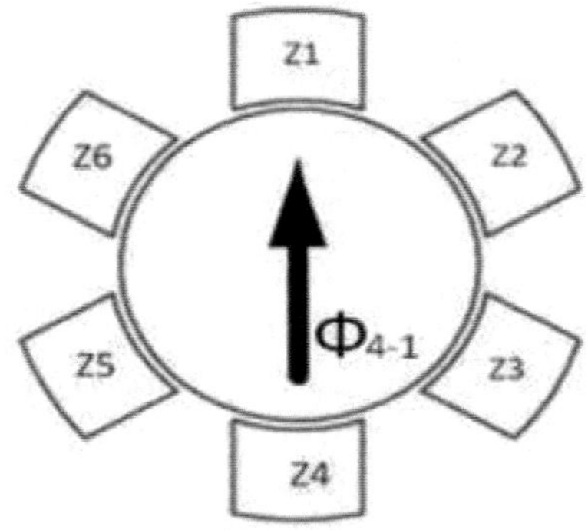

Figure 10. Resulting direction of the magnetic flux lines in the air gap from Z4 to Z1

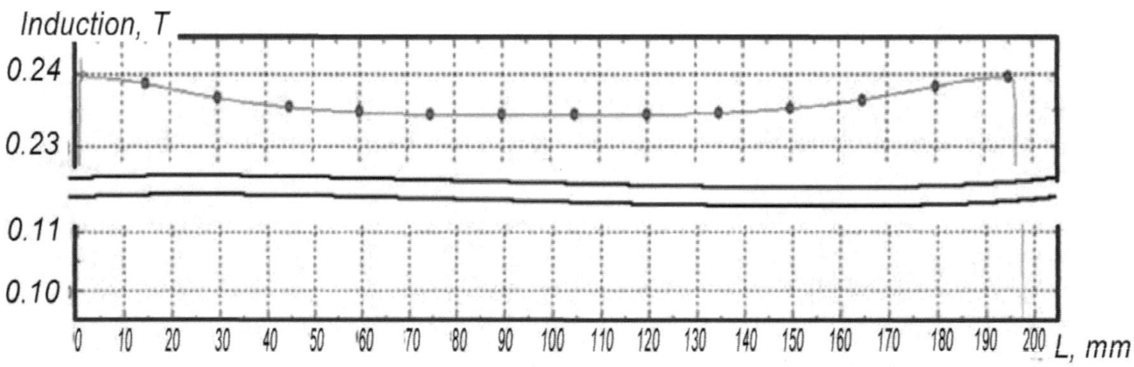

Figure 11. Induction change in the air gap along the rotor on the side of the tooth tips

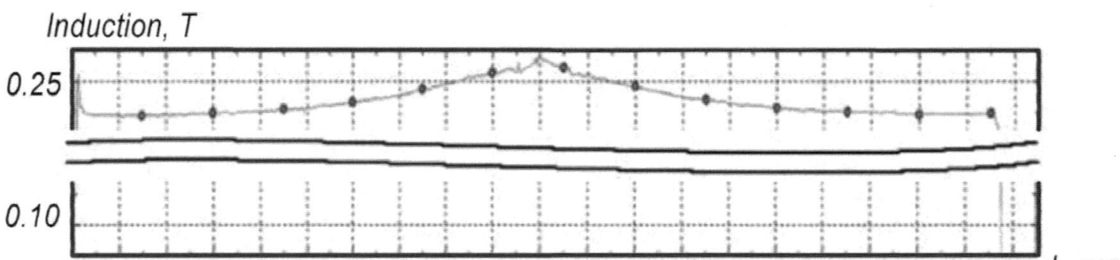

Figure 12. Induction change in the air gap along the rotor on the side of the tooth tips

Considering in a similar way the changes in inductions along the planes indicated in figures 8b and 8c, we obtain the flux directions according to figures 13 a, b.

At the indicated values of the phase currents relative to the rotor, the tooth tips Z6, Z1, Z2 create the North Pole. Accordingly, tooth tips Z5, Z4, Z3 create the South Pole. With an error of less than

IOP Conf. Series: Materials Science and Engineering **313** (2018) 012002 doi:10.1088/1757-899X/313/1/012002

2%, we can assume that the average values of the inductions under tooth tips Z6, Z2 are equal to each other. Also, with an error of less than 2%, we can assume that the average values of the inductions under tooth tips Z5, Z3 are equal to each other too.

When the phase current values change symmetrically according to sine law by an angle α in electric degrees, for example, by 60 electric degrees, the field symmetry axis is shifted by 60 geometric degrees, etc. When the phase sequence changes the rotation direction of rotation of the magnetic field changes too.

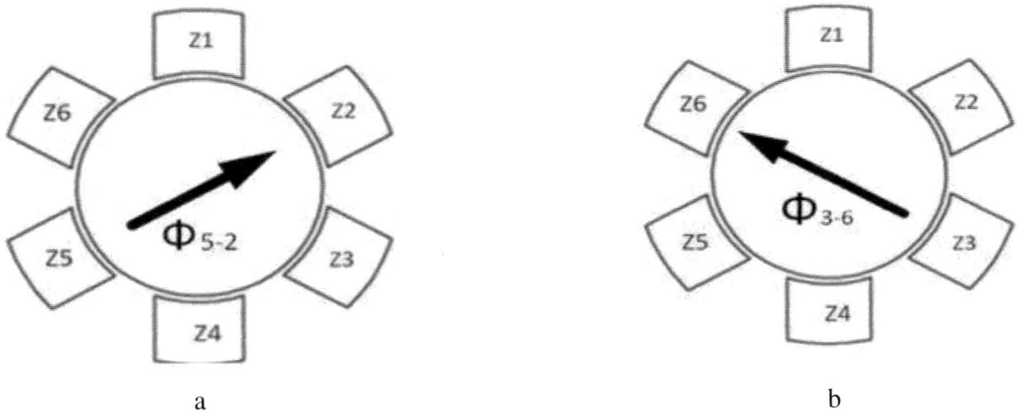

Figure 13. Resulting direction of the magnetic flux lines in the air gap:
a) From Z5 to Z2; b) From Z3 to Z6

Figure 14 shows a photograph of a laboratory sample performed for $m = 3$, $p = 5$, $q = 1$.

Figure 14. Laboratory sample of a motor with ring windings

The tests carried out due to the laboratory sample of the induction motor with ring windings have showed that:

1. The induction motor with ring windings has starting and maximum torques.
2. The direction of rotor rotation of the induction motor with ring windings does not depend on the starting angular position of the rotor. The change in the direction of rotation cans be obtained by changing the phase sequence of the supply voltage.

3. The synchronous rotational speed of the rotor for the prototype of the induction motor with ring windings is n = 600 rpm, which corresponds to a calculated number of pole pairs p=5.

4. Investigation of the harmonic composition of the magnetic field in the gap showed that the field in the gap has the same harmonic and amplitude series as for a classical machine with a classical winding with the number of slots per pole and phase q=1 and the diametric step.

5. Conclusion

1. With an error of less than 5%, we can assume that the considered principle of the formation of a periodically changing different-pole magnetic field moving relative to an imaginary cylindrical surface, which consists in a staged transforming the type of movement of the magnetic field of the armature with respect to an imaginary cylindrical surface (rotor) by means of a different configuration of magnetic masses and the distribution of coils of the armature winding, it is possible to create a symmetric field of mutual induction similar to the field created by the drum winding in induction motors of classical design.
2. The principle of making the number of pole pairs that is presented in [2] is valid. With the minimum possible number of coils of the stator winding equal to six and the number of phases m = 3, we can obtain any number of pole pairs.
3. Each stator coil of the induction motor with ring windings performs the function of the p-sides of the coils of the classical design machine.

References

[1] But D 1990 *Non-contact electrical machines* (Moscow: High School)
[2] Plastun A 1998 *Armature of polyphase electric machine* (Pat. 2121207: Russian Federation)
[3] Bauman E and Ivchenko Y 1981 *Rotor with claw-shape poles* (Pat. 873333: SSSR)
[4] Tikhonova O and Plastun A 2017 Electromagnetic Torque Calculation of Induction Motor with Ring Windings by dint of "ANSYS Maxwell *IEEE Conf. of Russian Young Researchers in Electrical and Electronic Engineering (Electronic materials, St. Petersburg: LETI)* pp. 251-255
[5] Alwodai A and Gu F 2012 Comparison of Different Techniques for Induction Motor Rotor Fault Diagnosis *Journal of Physics* pp. 1-11
[6] Thomson W T. and Fenger M 2001 Current signature analysis to detect induction motor faults *IEEE Industry Applications Magazine* pp 26–34
[7] Kopylov I 2002 *Electrical machines design* (Moscow: High School)
[8] Ben Sasi A Y, Gu F, Payne B and Ball A 2004 Instantaneous angular speed monitoring of electric motors *Journal of Quality in Maintenance Engineering* 10 pp 123–135
[9] Antik I 1938 *Inductuion motors winding data* (Moscow)
[10] Stavrou A, Sedding H G and Penman J 2001 Current Monitoring for Detecting Inter-turn Short Circuits in Induction Motors *IEEE Transactions on Energy Conversion* 16 pp 32–37

Study of the AC machines winding having fractional q

V Y Bespalov[1] and A O Sidorov[1]

[1] NRU MPEI, 111250, Krasnokazarmennaya 14, Moscow, Russia

E-mail: sao92@yandex.ru

Abstract. The winding schemes with a fractional numbers of slots per pole and phase q have been known and used for a long time. However, in the literature on the low-noise machines design there are not recommended to use. Nevertheless, fractional q windings have been realized in many applications of special AC electrical machines, allowing to improve their performance, including vibroacoustic one. This paper deals with harmonic analysis of windings having integer and fractional q in permanent magnet synchronous motors, a comparison of their characteristics is performed, frequencies of subharmonics are revealed. Optimal winding pitch design is found giving reduce the amplitudes of subharmonics. Distribution factors for subharmonics, fractional and high-order harmonics are calculated, results analysis is represented, allowing for giving recommendations how to calculate distribution factors for different harmonics when q is fractional.

1. Introduction

Modern electric drives with AC motors are often frequency controlled. Thanks to the invertor supply a machine has different speed regardless of a pole number $2p$, which in its turn is a result of optimal design. The best mass-dimension, say synchronous permanent magnet motors, have at $2p \geq 6{\div}10$ fractional q winding.

Fractional q is used first of all in limited size micromachines [1]. Also, it is good in powerful low speed hydrogenerators having large pole number, because great demands to dimensions and tooth saturation do not allow to use integer q winding. Besides, fractional q windings often have been used in the course of modernization – electric machines redesign to the lower rotation speed at the stator having odd slots number. When the new series RA of induction machines was under development [2, 3], symmetrical 3-phase fractional q winding has been applied in spite of their lower efficiency. It is worth mentioning quite successful use of fractional q (with $q < 1$) windings in reluctance machines thanks to linear load increase because of stator number of slots decrease [4].

Fractional q is one of the most effective mean to depress tooth-order harmonics. However, in the specialized literature on low-noise electrical machines, one can find recommendation to refrain from this windings application [5]. This is bound up, first of all, with high probability of low-order force waves development, which cause increased vibrations. This phenomenon is explained by appearance and interaction of harmonics with number of poles non-multiple to number of poles of the main harmonic $p_v \neq k\, p_1$, having orders less than main harmonic ones. Such harmonics have been called subharmonics or low-order ones. Besides of them, in vibration spectrum, there may be fractional and high-order harmonics.

IOP Conf. Series: Materials Science and Engineering **313** (2018) 012003 doi:10.1088/1757-899X/313/1/012003

2. Simulation and harmonic analysis

In order to reveal space harmonics in spectra of winding MMF and air gap flux density, a model was developed in FEMM program, which allows simulating a two-dimensional stationary magnetic field by the finite element method. Stator winding is made of copper, nonlinear magnetic core – of the steel 2412. Boundary conditions are: Dirichlet – on the outer bound of the machine active zone which is a stator outer diameter, Neumann – on all the rest bounds dividing mediums. Number of slots have been taken 36, $q = 1\frac{1}{5}$. Windings current density is given as

$$J = \frac{\sqrt{2} I_p w_s}{S_s} \tag{1}$$

where I_p – phase current, w_s – phase number of turns, S_s – stator slot area.

Phase currents are given for the time corresponding phase A maximum current, taking into account their directions of flow, that is $I_a = I_m$, $I_b = \frac{I_m}{2}$, $I_c = \frac{I_m}{2}$. Number of finite elements is equal to 250 thousands. Stator slots have trapeziform. To exclude slots influence rotor surface is considered to be smooth.

As a result of the simulation, flux density distribution B_δ is obtained along the imaginary line dividing the machine's air gap by two equal parts, for the integer and fractional q (Figure 1). As well seen, for both cases positive and negative half-waves have similar forms with the shift π radians. That is why flux density curve B_δ contains only odd harmonics, it is a second type symmetry.

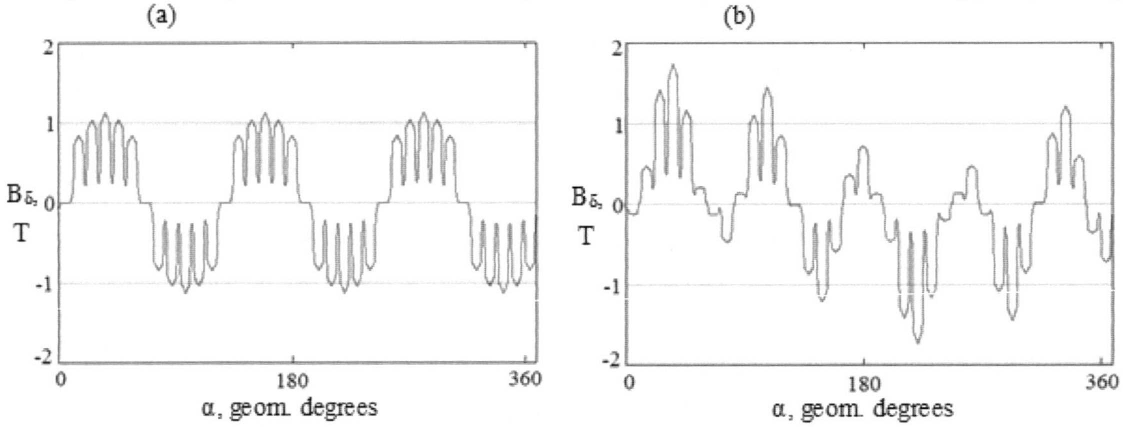

Figure 1. Flux density in the air gap of PM synchronous motor with: (a) integer q, (b) fractional q.

Fractional q windings possess two types of asymmetry. First, within each pole flux density curve is not symmetrical with respect to the pole axis, that is the first type symmetry is broken. The result is explained only by the fact that odd high-order harmonics intersect the abscissa axis not simultaneously with the main wave and no new harmonics are formed at the same time. Secondly, integral meanings of flux density B_δ (that is magnetic flux) are different under poles resulting different forces amplitudes acting on poles. This may give rise to the additional vibration components.

Results of field simulation are analyzed with Fourier series of field harmonic components (Figures 2a and 4a). Flux density in the gap is represented as the sum of terms of the harmonic series

$$B = \sum_{\nu=1}^{\infty} B_{\nu m} \cos \nu \frac{\pi}{T} x \tag{2}$$

where $B_{\nu m} = \dfrac{m\sqrt{2} I_p w_s \kappa_\nu \mu_0 \lambda_\delta}{\pi \nu p}$; T - pole pitch of the main space harmonic, m; ν - space harmonic order; x - space coordinate, m; m - phases number; k_ν – winding factor; $\mu_0 = 4\pi 10^{-7}$, H/m – space permeability; p - number of pole pairs of fundamental harmonic; λ_δ - air gap specific magnetic permeance.

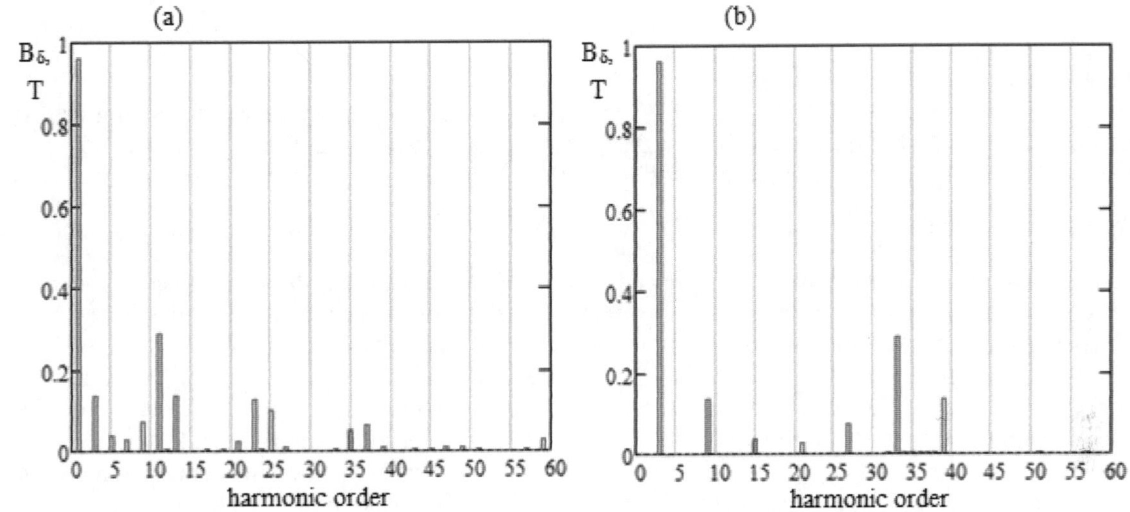

Figure 2. Flux density harmonics of induction motor with integer q: (a) $T = 2\pi$, (b) $T = 6\pi$.

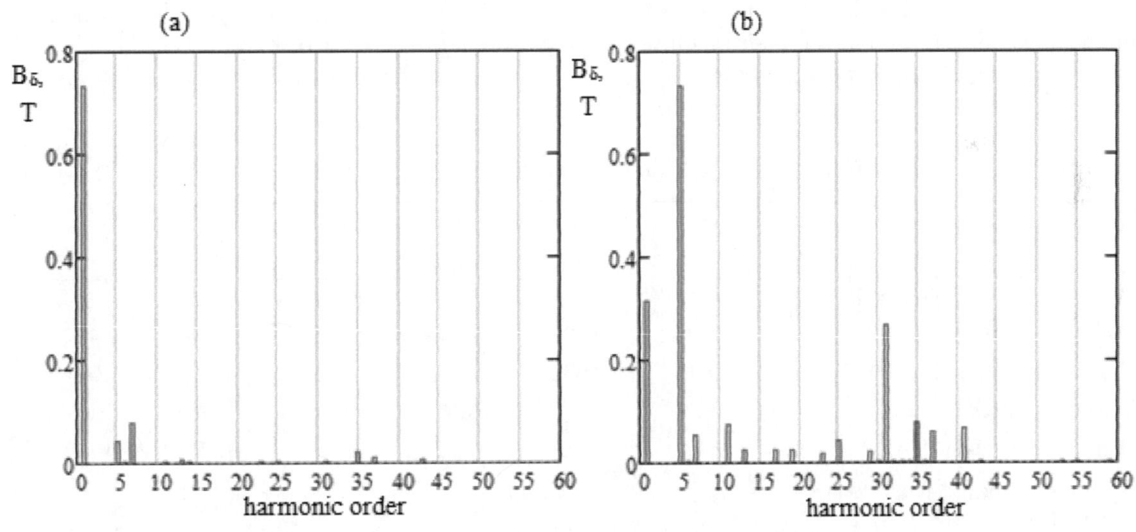

Figure 3. Flux density harmonics of induction motor with fractional q: a) $T = 2\pi$, b) $T = 6\pi$.

In this case, the period of the fundamental harmonic is equal to $T = 2\pi$. When considering subharmonics, their period will be c times greater $T_\nu = cT$. Therefore, it is possible to identify subharmonics by increasing the period of the fundamental harmonic by a factor of c (Figures 2b and 3b).

The spectrum of the harmonics of the machine, as expected, does not contain even harmonics by virtue of the symmetry of the flux density curve in the air gap relative to the abscissa axis under the shift of half waves.

IOP Conf. Series: Materials Science and Engineering **313** (2018) 012003 doi:10.1088/1757-899X/313/1/012003

The non-tooth orders caused by the nonuniformity of the distribution of the MMF curve and the different air gap permeance are present in the windings with integer and fractional q:

$$\nu = \frac{2m}{d_q}k + 1 \tag{3}$$

where $k = \pm1; \pm2\ldots$

There is a wide range of flux density tooth-order harmonics in machine with integer q:

$$\nu = \frac{Z_1}{p}k + 1 \tag{4}$$

where $k = \pm1; \pm2\ldots$

For our example ν = 11, 13, 23, 25, 35, 37, 47, 49… In the flux density curve of winding having fractional q, these harmonic components disappear or considerably decline. However, if main harmonic period increase, one may note subharmonic components and that maximal amplitude has harmonic of order $\nu = \frac{1}{5}$. In spite of the low order, its amplitude reaches a value commensurable to the main component. Namely this harmonic causes flux density asymmetry between poles in the gape and consequently unequal distribution of electromagnetic forces acting on the machine rotor. Subharmonic effect can be brought down through the rational choice of winding pitch y.

Air gap field spectra for different winding pitch y are shown in the table 1. Subharmonic amplitude may reach 60% at $y = 6$. As can be seen, the amplitudes of the subharmonics for different y acquire different values. For a given winding layout, it is possible to reduce subharmonics thanks to the choice of proper y. In our calculation y is varied from 1 to 6. For $y = 1$ and $y = 6$ main harmonic amplitude goes down considerably but subharmonics and high-order harmonics amplitude grow. Fundamental amplitude maximum takes place at $y = 3$ and $y = 4$. In addition, their characteristics are better than at the rest variants. Within the considered range, an optimal harmonics minimum is at $y = 3$. It means that if we design a low-noise machine, one must consider not only influence of high-order harmonics, but subharmonics as well, since their amplitudes may reach high values. However, it should be noted if some harmonics are to be suppressed, it may be possible to choose certain interim y in order to reduce undesirable components of the spectrum.

Table 1. Air gap field spectra for different winding pitch y.

	winding pitch y					
	6	5	4	3	2	1
Main harmonic amplitude, T	0.41	0.63	0.74	0.74	0.62	0.35
Amplitude of harmonic 1/5, T	0.248	0.193	0.149	0.122	0.105	0.067
Amplitude of harmonic 3/5, T	0.01	0.012	0.007	0.02	0.002	0.001
Subharmonic factor, %	7.2	10.1	8.4	6.4	9.2	12.7
High harmonic factor, %	61.1	30.7	20.2	16.8	16.9	19.2
Common factor of subharmonics and high-order harmonics, %	61.6	32.3	21.9	18	19.3	23

3. Frequencies of electromagnetic forces

Electromagnetic vibration component mostly is calculated by a simple Maxwell formula:

$$P = \frac{B_\delta^2}{2\mu_0} \tag{5}$$

It allows to find a radial force acting on the rotor and shaft of electrical machine. Axial and tangential components are ignored. Axial force appears only because of eccentricity and considerable defects of magnetic system, such as fluffed up rotor or stator laminated core. Tangential component causes vibrations of magnetic core teeth. Their natural frequency of the teeth usually exceeds 10 kHz, but electromagnetic forces have frequencies limited by 4 kHz. That is why only radial component of force is taken into account.

Air gap flux density is a sum of components: created by motionless stator (or armature) B_s and rotating rotor (or inductor) B_r. In this way:

$$B_\delta = B_s + B_r \tag{6}$$

Each of them can be represented by a product of MMF and magnetic permeance:

$$B_\delta = F \cdot \Lambda \tag{7}$$

Space harmonics in a permanent magnet synchronous motor rotate together with rotor at a synchronous speed $n_\mu = n_1$. High harmonics have number of poles with respect to the fundamental harmonic $p_\mu = \mu \cdot p_1$. Therefore harmonics induce EMF with a frequency in the stator:

$$f_\mu = \frac{p_\mu n_\mu}{60} = \frac{\mu \cdot p \cdot n_1}{60} = \mu \cdot f_1 \tag{8}$$

Stator current also create space harmonics of the stator winding. Their number of poles, like rotor's ones, is equal to the fundamental, but speed is lower. Thus, the frequency of EMF from their high harmonics is equal:

$$f_v = \frac{p_v n_v}{60} = \frac{v \cdot p_1 \cdot n_1 / v}{60} = f_1 \tag{9}$$

It is known that pulsating MMF wave can be represented as a sum of the forward and backward rotating waves. Taking into consideration high-order harmonics, one may have an expression for the phase MMF:

$$F = \sum_{v=1}^{\infty} F_m \sin \omega t \cos v \frac{\pi}{\tau} x = \sum_{v=1}^{\infty} \frac{F_m}{2} \left[\sin\left(\omega t - v\frac{\pi}{\tau}x\right) + \sin\left(\omega t + v\frac{\pi}{\tau}x\right) \right] \tag{10}$$

Having found each flux density component, one may write the radial magnetic force as follows:

$$P = \frac{1}{2\mu_0} \left[\left(\sum_{v=1}^{\infty} F_v \sin\left(\omega t - v\frac{\pi}{\tau}x\right) + \sum_{\mu=1}^{\infty} F_\mu \sin\left(\mu\omega t - \mu\frac{\pi}{\tau}x\right) \right) \times \left(\Lambda_0 + \sum_{v=1}^{\infty} \Lambda_m \cos vZ_1 \frac{\pi}{\tau} x \right) \right]^2 \tag{11}$$

Let's raise to the second power the first multiplier. Since we consider interaction of stator and rotor harmonics, this expression will be:

$$P = \frac{1}{2\mu_0} \left[\left(\sum_{v=1}^{\infty} F_v \sum_{\mu=1}^{\infty} F_\mu \sin\left(\omega t - v\frac{\pi}{\tau}x\right) \sin\left(\mu\omega t - \mu\frac{\pi}{\tau}x\right) \right) \times \left(\Lambda_0 + \sum_{v=1}^{\infty} \Lambda_m \cos vZ_1 \frac{\pi}{\tau} x \right)^2 \right] \tag{12}$$

Using the sines product formula we finally get

$$P = \frac{1}{4\mu_0}\left[\left(\sum_{\nu=1}^{\infty} F_\nu \sum_{\mu=1}^{\infty} F_\mu \cos\left((1-\mu)\omega t - (\nu-\mu)\frac{\pi}{\tau}x\right) - \cos\left((1+\mu)\omega t - (\nu+\mu)\frac{\pi}{\tau}x\right)\right) \times \left(\Lambda_0 + \sum_{\nu=1}^{\infty} \Lambda_m \cos \nu Z_1 \frac{\pi}{\tau}x\right)^2\right] \quad (13)$$

Thus, electromagnetic forces frequencies in permanent magnet synchronous motor will be

$$f = f_\nu \pm f_\mu = f_1 \pm \mu \cdot f_1 = f_1 \pm (1+2k)f_1 = \begin{bmatrix} 2\cdot(1+k)f_1 \\ 2kf_1 \end{bmatrix} \quad (14)$$

As it is seen, for machines with a field winding and with permanent magnet, frequencies coincide. The same result one can get substituting $Z_2 = 2p$ and $s = 0$ for induction machine formulas ignoring saturation:

$$f = f_\nu \pm f_\mu = f_1 \pm \mu \cdot f_1 = f_1 \pm \left(1 + \frac{(1-s)Z_2}{p}\right)f_1 = \begin{bmatrix} \frac{(1-s)Z_2}{p} f_1 \\ \left(2 + \frac{(1-s)Z_2}{p}\right) f_1 \end{bmatrix} \quad (15)$$

Pay attention at subharmonics which appear mainly at double supply frequency and divisible by its frequencies.

4. Winding factor

It is one of the main criteria for choice and design of the electrical machine winding. Winding factor reflects the influence of high-order harmonics and consists of three components:

$$k_\nu = k_{p\nu} k_{d\nu} k_{sk\nu} \quad (16)$$

where $k_{p\nu}$ - pitch factor, $k_{d\nu}$ - distribution factor, $k_{sk\nu}$ - skewing factor.

An effective way to suppress high-order harmonics (first of all, tooth-order harmonics) is a slot (teeth) skewing on the stator or rotor. However, some problems arise when realizing this measure. To skew rotor magnets, one must make them segmental, it is necessary to find a way of their mounting (fixing) that is rotor assembling gets profoundly complicated. To make stator slots skewed it is not always possible as well. Therefore, application of the fractional q becomes a priority mean to do away with high-order harmonics. For this reason we take here $k_{sk\nu} = 1$. Pitch factor for each harmonic is given:

$$k_{y\nu} = \sin\left(\nu\beta\frac{\pi}{2}\right) \quad (17)$$

where β - relative winding pitch.

As a rule, pitch factor is not considered in the analysis of fractional q windings, since parameter q itself does affect this coefficient. However, it is worth mentioning that the winding pitch choice may considerable influence on a subharmonic amplitude, that is to be taken into account in the design procedure of low-noise electrical machines.

In the paper, analysis of distribution factor $k_{d\nu}$ for a main, high-order and subharmonics. Several known formulas for $k_{d\nu}$ are as follows.

Method [6] for integer and fractional ν is:

$$k_{p\nu} = \frac{\sin(\frac{\pi}{2m}\nu)}{C_q \sin(\frac{\pi}{2mC_q}\nu)} \tag{18}$$

Method [7] for integer ν is:

$$k_{p\nu} = \frac{\sin(\nu\frac{\pi}{k_z m})}{C_q \sin(\nu\frac{\pi}{k_z mq})} \tag{19}$$

Method [7] for fractional ν is:

$$k_{p\nu} = \frac{\sin(\nu\frac{\pi}{k_z m} d_q y_i)}{C_q \sin(\nu\frac{\pi}{k_z mC_q} d_q y_i)} \tag{20}$$

Method [8] for integer ν is:

$$k_{p\nu} = \frac{\sin(\nu \cdot 30)}{d_q q \sin(\nu\frac{30}{d_q q})} \tag{21}$$

Method [8] for fractional ν is:

$$k_{p\nu} = \frac{\sin(\nu \cdot 30 - \nu \cdot 180)}{d_q q \sin(\nu\frac{30}{d_q q} \mp \nu \cdot 180)} \tag{22}$$

Method [9] for integer and fractional ν is:

$$k_{p\nu} = \frac{\sum \cos n\varphi}{2p} \tag{23}$$

Here in k_z – zone coefficient; C_q - numerator of the improper fraction q; d_q – denominator of the improper fraction q; φ - angular coordinate of the phase conductor; n - number of conductors having the same coordinate.

Figures 4 and 5 represent distribution factor calculated by these formulas for $q = 1\frac{1}{5}$. According to the theorem on periodicity of pitch factors they will be repeated periodically. Therefore it is quite enough to consider them within a limited range. It is seen that distribution factor meanings found by formulas [6-8] coincide for high-order harmonics. Only meaning found by [9] differ, since in them all even harmonics turn into zero.

Pay attention to results of calculation of the distribution factor for subharmonics. Formulas [6-9] give different $k_{d\nu}$, that can be explained as follows. Method [6] is based on the assumption that fractional q is substituted by a meaning calculated for an equivalent number of slots, what does not allow to calculate distribution factor for subharmonics and fractional-order harmonics. Besides, while calculating by methods [7, 8], even harmonics get nonzero meanings. As a matter of fact it is not quite correct result. In spite of fractional q and presence of subharmonic components, a winding is symmetrical and even harmonics must not be present in it. Nonzero meanings occur because, substituting q in formulas, its meaning changes from $q = 1\frac{1}{5}$ to $q = 1\frac{2}{10}$, that is number of elementary periods becomes twice as much and winding characteristics change. As a result winding gets unsymmetrical features and distribution factors k_{dv} for subharmonics and fractional-order harmonics become nonzero.

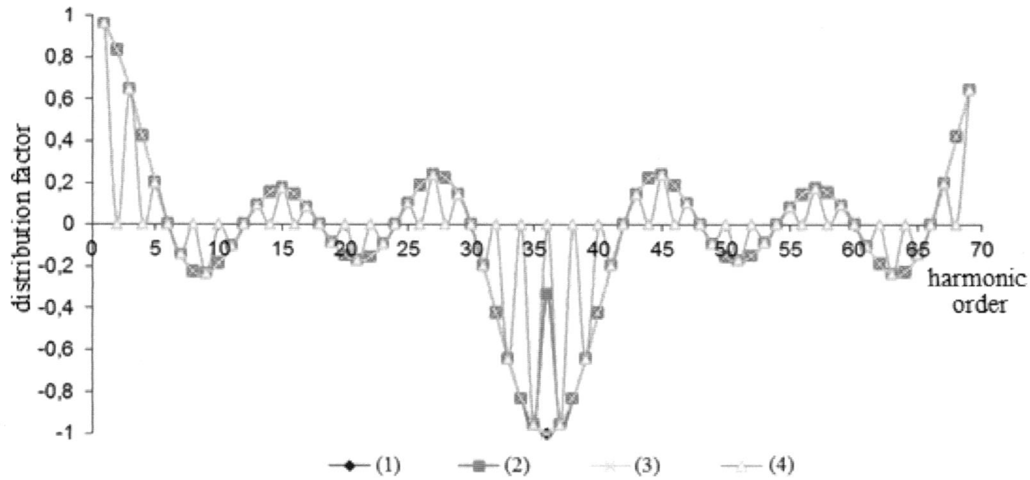

Figure 4. Distribution factor for high-order harmonics.

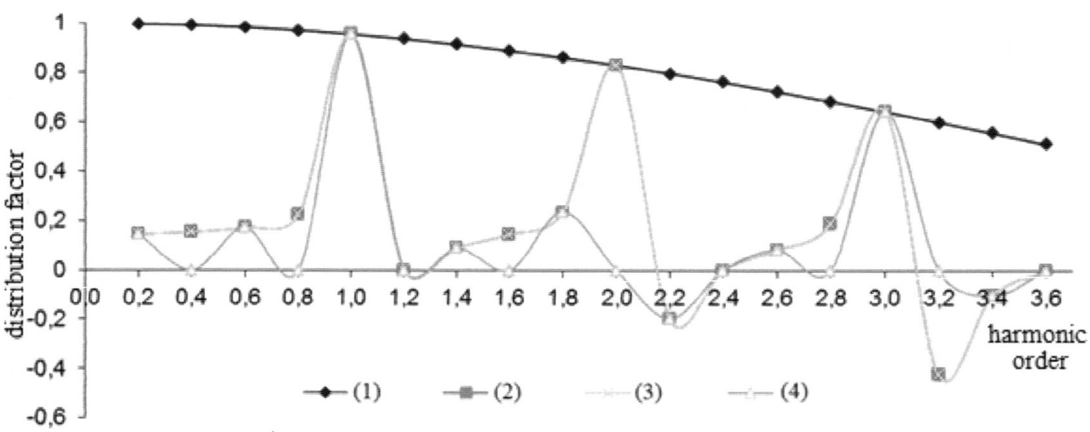

Figure 5. Distribution factor for subharmonics and fractional harmonics.

Method [9] allows to calculate k_{dv} on the base of space EMF distribution and angular conductor positions φ. Thanks to them, it is easy to calculate distribution factor for subharmonics and fractional components correctly. These data is confirmed by results of flux density harmonics spectra calculations:

$$B_{\nu} = \frac{1}{\nu}\frac{I_1}{I_0}\frac{k_{\nu}}{k_1}k_s \tag{24}$$

For simplicity we ignor saturation k_s and consider no load operation, that is $I_1 = I_0$.

On Figure 6, we noted flux density subharmonics received by winding factor calculations with different formulas.

Comparison of calculation and simulation results (Figure 2b) reveals better convergence of the method [9]. For experimental verification of these results, it is planned to produce permanent magnet synchronous motor with $q = 1\frac{1}{5}$ model, operating at subharmonic frequency $2f_1$.

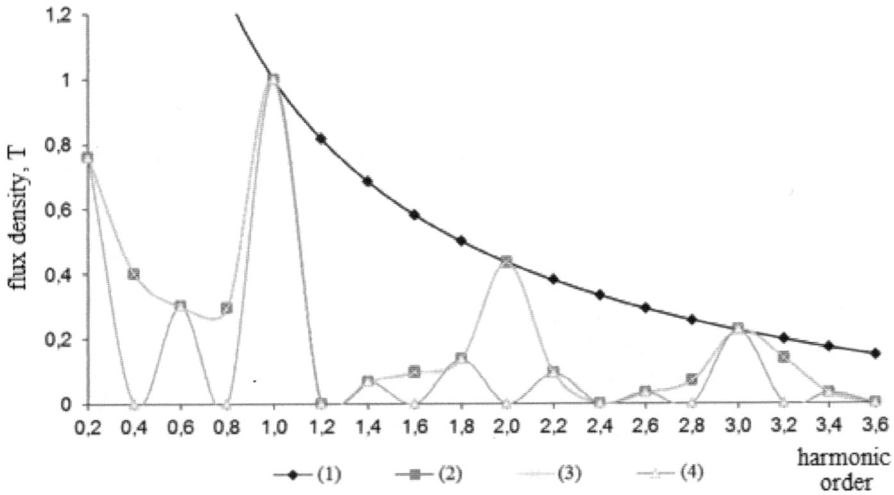

Figure 6. Calculated flux density in air gap of subharmonics.

5. Conclusions

1. Flux density harmonics spectrum in the air gap of permanent magnet synchronous machine, when fractional q is used, contains subharmonics and fractional harmonics which may cause heightened vibrations on certain harmonics.
2. At different values of winding pitch y, subharmonic amplitudes will be different, that must be taken into account in the design procedure of low-noise machines.
3. Frequencies of subharmonics and high-order harmonics coincide with frequencies of electromagnetic vibration sources for salient pole synchronous machines and are equal to $2fk$, where k=1,2,3,…
4. Calculation methods [6-9] of distribution factor k_{dv} differ for subharmonics and fractional harmonics. Method [9] provides the best meeting with results of simulation approach.

References

[1] Volkov L K 1979 Vibration and Noise of Small Electrical Machines ed V A Romanov (Leningrad: Energy)
[2] Popov V I, Ahunov T A and Makarov L N 1999 *Modern Induction Eectrical Machines: New Russian series RA* (Moscow: Znack)
[3] Goldberg O D, Makarov L N and Khelemskaya S P 2016 Engineering Design of Electrical Machines (Moscow: BASTET)
[4] Shevchenko A F and Chestyunina T V 2009 *J. Elektrotechnika* **12** pp 3 – 7
[5] Shubov I G 1973 *Noise and Vibration of Electrical Machines* ed V N Mikhankova (Leningrad: Energy)
[6] Kopylov I P 1980 *Electric Machinery Design* (Moscow: Energy)
[7] Voronkin V A, Gecha V Ya, Gorodetskiy E A, Yevlanov V V, Zakharenko A B, Zubrenkov B I, Kaplin A I, Ledovskoy V I, Sklyarova I V, Smirnova L P and Shapiro M Kh 2006 *J. Electromechanical Matters* **103** p 178
[8] Astakhov N V and Yurgenson T S 1987 *Magnetic vibrations of synchronous machines* ed V T Tikhonov (Moscow: MPEI)
[9] Jerve G K 1989 *Electrical Machinery Windings* ed L M Parkhomenko (Leningrad: Energoatomizdat)

IOP Conf. Series: Materials Science and Engineering **313** (2018) 012004 doi:10.1088/1757-899X/313/1/012004

Variable cross-section windings for efficiency improvement of electric machines

P Yu Grachev[1], A A Bazarov[1] and A S Tabachinskiy[2]
[1]Professor, Samara State Technical University, Samara, Russian Federation.
[2]PhD student, Samara State Technical University, Samara, Russian Federation.

Tabachinski.as@ieee.org

Abstract. Implementation of energy-saving technologies in industry is impossible without efficiency improvement of electric machines. The article considers the ways of efficiency improvement and mass and dimensions reduction of electric machines with electronic control. Features of compact winding design for stators and armatures are described. Influence of compact winding on thermal and electrical process is given. Finite element method was used in computer simulation.

1. Introduction

The serious problem of efficiency improvement of electromechanical converters is often solved by implementing advanced control algorithms, which are able to afford process quality without any changes in power converter construction [1].

In the Ural region, some researches of construction development for electric machines were made recently [2,3]. The aim of the researches was to improve efficiency and reduce mass and dimensions.

For vehicles application, in Japan special stator construction of starter-generator was proposed [4]. This construction includes rectangular winding wires that are able to reduce overhang deviation and copper intensity.

Even greater efficiency improvement and dimensions reduction of autonomous power units, hybrid (HEVs) and electric vehicles (EVs) in particular, can be reached by multipolar machine and multiphase electronic converter integration. In Germany, alternating current machine with short-circuited stator winding and integrated electronic converter is approved [5]. Electronic converter can change virtual number of machine phases during the operation, which leads to increased output performance of power unit. Extra-low operating voltage is the disadvantage of this AC machine.

The researches of integrated power units for hybrid vehicles, wind turbines and micro-hydro are holding in Samara [6]. The AC machine construction with compact windings leads to significant reduction and overhang deviation, mass and dimensions, both with efficiency improvement. The main idea of winding construction is in periodical variation of rectangular conductor cross-section [7,8,9]; this construction allows varying air gap between stator core and overhang parts, as well as overhang deviation.

In this article, construction features of compact windings for DC and AC machines are described. In windings with irregular cross-section irregular current density is observed [10]. The results of computer simulation of electromagnetic and thermal fields that are given in the article allow to measure phase resistance and temperature in active and overhang parts, as well as to design material-saving construction [11]. Construction features of compact winding can be used in other types of electric machines [12].

2. Machine stator with irregular cross-section winding

Let us consider construction features of this winding. On figure 1 stator with three-phase rectangular concentrated winding and series connected branches is illustrated. Winding phase consists of three branches. Overhang parts of compact winding are parallel to stator core end-faces. There are jumpers between overhang and active conductors, and jumper cross-section is less than others. This allows to place overhang parts in the same plane without overlapping [8].

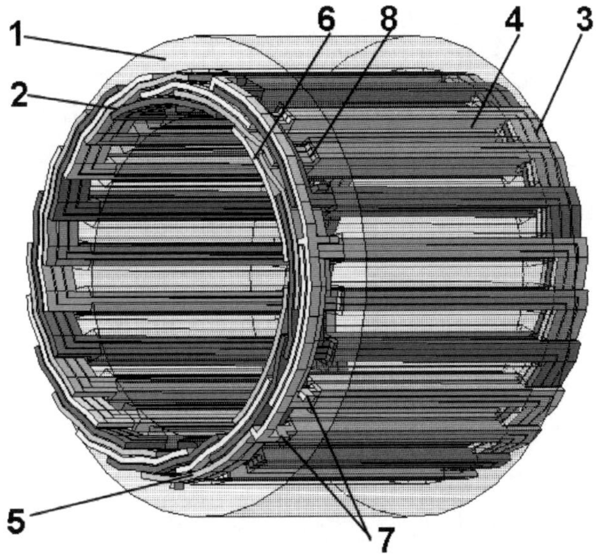

Figure 1. Electric machine stator with compact winding.

On figure 1 we can see: 1 — stator core, 2 — winding, including active parts 3 and overhang parts 4. Overhang conductors 5, placed near the core yoke, connect branches of the winding phases. Overhang conductors 6 near the air gap connect half-branches. Phasing outs 7 are connected with active conductors of the bottom layer. Cross-section of jumpers 8 that connect active and overhang conductors is double reduced. This allows to significantly decreasing the overhang deviation.

From the other end-face there are no phasing outs and branches connection. Overhang construction does not depend on winding scheme and branches connection. Jumpers from this side has also reduced cross-section and the deviation is also decreased.

Due to winding length reduction, active power losses are decreasing and efficiency is increasing. Volume reduction is extremely important for autonomous application, like electric vehicles and wind turbines, because of the strict dimensions requirements. Nevertheless, for new form of overhang conductors, which allow to vary air gap between stator core and overhang conductors, new electrical and thermal processes should be considered [11]. It is very important to research the dynamic processes of compact winding machine, for example, for start processes of internal combustion engine of hybrid vehicles [13].

3. New winding for DC machine armature

On figure 2 composition of brushed DC machine with new armature winding construction [14]. In the housing 1 stator 2 is placed, generating the stationary magnetic field. Stator consists of yoke and pole with field windings. Armature 3 divided from stator 2 by air gap 4, and consists of armature core 5 with slots 6 (marked by dashed line). Armature is placed on shaft 7 and rotating on the bearings 8. Armature winding has two layers and includes overhang conductors 9 and 10 from both end-faces and active conductors 11. Outer parts of overhang conductors 10 play the role of slip rings 12 of commutator. Commutator also includes carbon brushes 13. Overhang insulation 14 also insulates slip rings of the commutator. Air gap 15 improves air convection for better cooling of overhang parts.

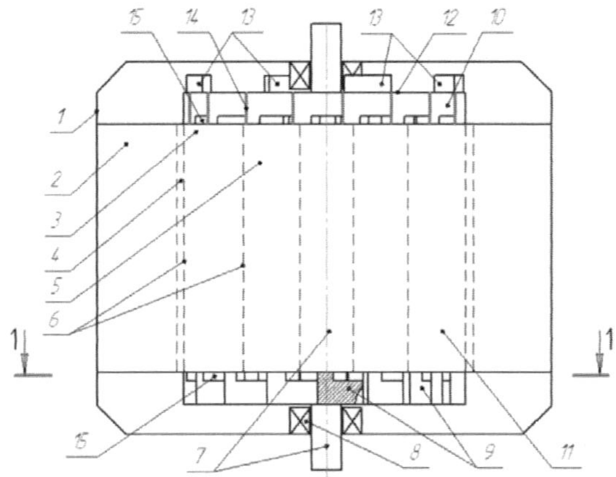

Figure 2. Composition of DC machine with compact armature winding.

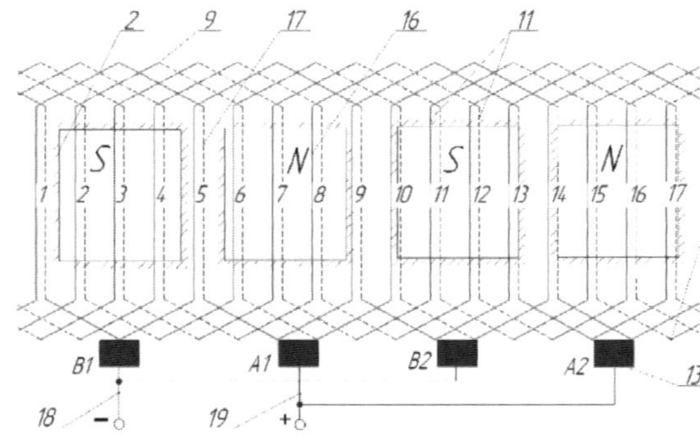

Figure 3. Winding scheme of armature.

On figure 3 armature winding scheme for this DC machine is illustrated. Wave bar winding has number of poles $2p=4$, $Z=Z_E=S=K=17$, $y1=4$, $y2=4$, $yk=8$. Stator poles marked with N and S. Solid lines illustrate active conductors of the upper layer 16, and dashed lines — bottom layer 17. We can also see overhang conductors 9 and 10. Brushes A1, A2, B1 and B2 connected to power supple bus 18 and 19.

On figure 4 top view of DC machine armature is illustrated. On figure 5 cross-section 1-1 from figure 2 is illustrated.

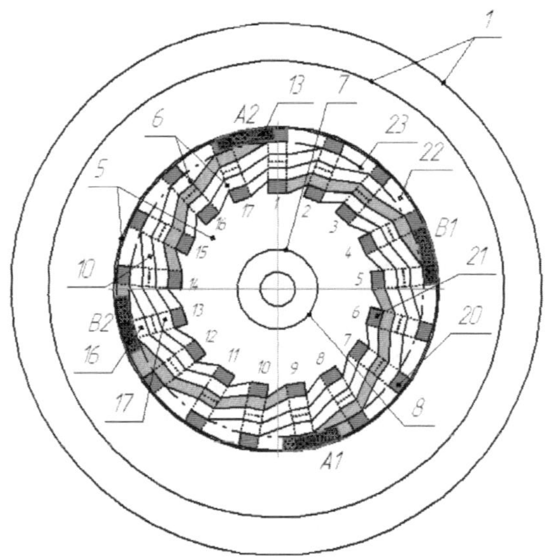

Figure 4. End-face composition of considered DC machine.

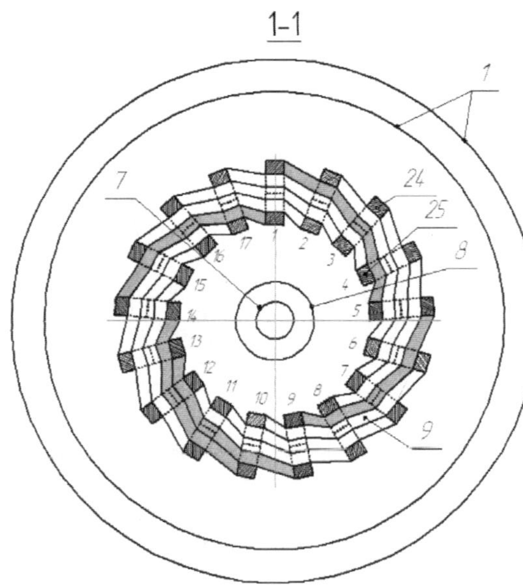

Figure 5. End-face cross-section of considered DC machine.

IOP Conf. Series: Materials Science and Engineering **313** (2018) 012004 doi:10.1088/1757-899X/313/1/012004

Housing 1 is transparent. Rectangular slots 6 of armature 5 are numbered clockwise. In the slots active conductors (bars) of upper 16 and bottom 17 layer are placed. They are connected to each other by overhang conductors 10. As we can see, jumpers 20, 21 between active and overhang conductors have reduced cross-section (marked with grey) — for this particular case cross-section of jumpers is 2/5 of active conductor cross-section. Jumpers are placed on the outer sides of active armature zone. This zone includes slots with active conductors and teeth and was described by Inkin A.I. in [13].

On the figure 4 we can see, that a part of each overhang conductor 10 is made like slip ring 12 with smooth surface 22 for commutator. Slip ring is bordered by dashed circle 23 and outer side of overhang parts 10. Brushes 13 are marked with hatched area.

Construction of overhang parts 9 and 10 is equal. On figure 5 jumpers 24 and 25, connecting active conductors with overhang conductors, are illustrated.

Proposed brushed DC machine operates like serial brushed DC machine with stationary stator.

Thus, new patented compact construction of DC machine allows to replace typical commutator with integrated slip ring the overhang conductor surface. This leads to mass and dimensions reduction and improvement of efficiency.

4. Electrical processes in the winding

Overhang parts shape, made like slip rings over the stator core end-faces, imposes the use of Ohm's law in integral and differential form [15]. Experimental [16,17] and theoretical [18] researches of current flow in flat conductors were made. Complex construction of winding turns features calculations of overhang deviation and phase winding parameters.

Calculation of phase resistance considers irregular current density for reduced cross-section conductors. On figure 6 finite element method was used in ELCUT 5.1 for current density simulation of compact winding.

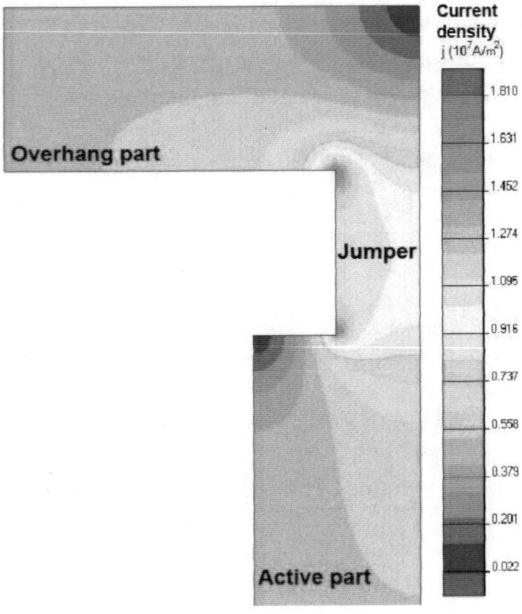

Figure 6. Current density in compact armature winding.

Results of simulation have shown that irregular current density influences resistance calculation only for short active zone length (*l<0.5D*). For longer active zones simplified methodic of phase resistance calculation is proposed [10]. In case of serial machine design, irregular current density consideration specifies resistance calculation for less than a percent.

IOP Conf. Series: Materials Science and Engineering **313** (2018) 012004 doi:10.1088/1757-899X/313/1/012004

5. Efficiency calculation

From the figure 1 for another 24-slot stator of AC machine we can see 4 overhang conductors over the slot area, two of them are placed in the free space and two of them are connected by jumpers with active conductors. Thus, for this particular winding overhang conductor width can be calculated:

$$b_O = \frac{h_S - 2h_J}{n} \tag{1}$$

here: b_O – overhang conductors width with insulation; h_S – height of all active conductors in the slot with insulation; h_J – jumpers height (in this case half of active conductor height); n – number of overhang conductors over one slot (depends on coil span).

Height of overhang conductor, in case of cross-section equality with active conductors, is:

$$h_O = \frac{S_S}{b_O} \tag{2}$$

here: h_O — overhang conductor height with insulation, S_S — slot conductor cross-section.

On the figure 1 three overhang conductors in a group, one over another, could be seen. Number of branches in a phase defines number of overhang conductors in a group. For this particular case, overhang deviation with air gap Δ is:

$$l_{OD} = 3h_O + \Delta \tag{3}$$

Calculations have shown the advantages of compact winding construction: for 80mm shaft height generator 4AN80A8U3 with rated voltage 220V, overhang deviation reduces from 29.8 to 12.7mm, active core length reduces by 20%, copper intensity reduces by 16.2%, and efficiency rises from 92% to 94%.

6. Thermal processes in the machine

For thermal flow calculation in stator the next 3D mathematical model is given:

$$\frac{\partial T(r,x,t)}{\partial t} = \nabla \cdot (k\nabla T) + \frac{1}{c\gamma} Q(x,y,z,t) \tag{4}$$

with border an initial conditions:

$$\begin{cases} T(x,y,z,0) = T_0 = T_c; \\ n \cdot (k\nabla T) = \alpha(T - T_C); \end{cases} \tag{5}$$

here: n – normal vector to the surface of computational area; c – specific heat capacity of steel, conductors and insulation; γ – density of steel, conductors and insulation; α – heat transfer coefficient; k – thermal conductivity of steel, conductors and insulation; $Q(x,y,z,t)$ - distribution function of inner heat sources.

Finite element method was used to calculate temperature distribution in stator core and active conductors for different air gaps between stator core and overhang conductors [10,13]. One case of temperature distribution is illustrated on figure 7.

For small air gap between stator core and overhang parts, temperature of overhang conductors rises. In every group of overhang conductors thermal condition of different levels varies. Overhang conductors can have common insulation for better heat transfer. Air gap needs to be wider for better air convection and heat transfer between overhang parts and stator active zone.

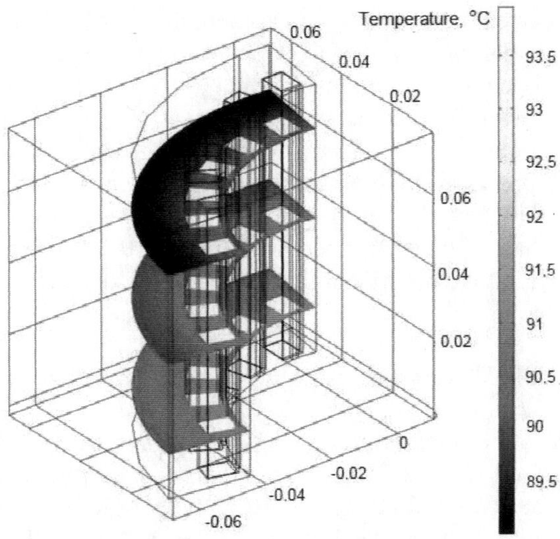

Figure 7. Stator core temperature distribution for 60A phase current and 2mm air gap.

Temperature distribution for wider air gap in active and overhang parts was researched. The results are given in table 1. This results allow to consider that wider air gap strives to equality of active and overhang conductor temperatures. But overhang deviation also increases.

Other cooling systems for these machines are scheduled to research. Frequency rotation control [20,21] and vibration consideration [22], for example.

Table 1. Temperature of active and overhang conductors.

№	Heat transfer coefficient, $W/(m^2 {}^\circ C)$		Active conductor temperature, $^\circ C$		Overhang conductor temperature, $^\circ C$	
	Stator end-face	Overhang parts	min	max	min	max
1	30	29	107	109	108	109
2	30	35	98.8	101	99.1	100
3	30	39	86.8	88.9	86.8	87.7
4	60	43	73.5	75.8	72.8	73.6
5	60	29	90.3	92	89.8	90.3

Conclusions

1. Proposed compact winding construction for DC and AC machines with periodical variation of conductor cross-section allows to decrease overhang deviation, reduce copper intensity and improve efficiency.

IOP Conf. Series: Materials Science and Engineering **313** (2018) 012004 doi:10.1088/1757-899X/313/1/012004

2. Research of irregular current density in compact winding was held via finite element method and computer simulation. The results have shown negligible effect of irregular current density on phase resistance and temperature of conductors and insulation.

3. Methodic of machine design with compact winding was given. The methodic considers features of compact winding construction.

4. Research of thermal processes in compact winding have shown the necessity of common insulation for overhang conductors for temperature gradient reduction. For improvement of heat transfer and decrease of overhang temperature air gap between stator core and overhang parts needs be increased. These decisions will make possible to increase maximum power of the machine.

References

[1] Seryi I M, Nosyrev M B and Ivanushkin B A 2008 Automatic control features of traction electric drive *Mining Journal* vol 8 pp 132-135.

[2] Antipin A S, Frizen V E and Sarapulov F N 2017 Power and Energy Parameters of the Inductor with an Electromagnetic Shield *15th International Conference on Electrical Machines, Drives and Power Systems (ELMA)*; DOI: 10.1109/ELMA.2017.7955473.

[3] Malygin I V and Plastun A T 2016 Synchronous Motor with Ring Windings Electromagnetic Torque Calculation Algorithm with ANSYS Package Usage *2nd International Conference on Industrial Engineering, Applications and Manufacturing (ICIEAM);* DOI: 10.1109/ICIEAM.2016.7911487.

[4] Ishigami T, Tanaka Y and Homma H 2015 Motor stator with thick rectangular wire lap winding for HEVs *IEEE Transactions on Industry Applications* vol 51 issue 4 pp 2917-2923.

[5] Patzak A, Bachheibl F, Baumgardt A, Dajaku G, Moroz O and Gerling D 2016 ISCAD – Electric High Performance Drive for Individual Mobility at Extra-Low Voltages *SAE Int. J. Alt. Power* vol 5 pp 148-156

[6] Grachev P Yu, Gorbachev E E and Tabachinskiy A S 2016 Power Converters with Compact Winding SCIGs for HEVs, Micro Hydroelectric Power Stations and Wind Turbines *2nd International Conference on Industrial Engineering, Applications and Manufacturing (ICIEAM);* DOI: 10.1109/ICIEAM.2016.7911521.

[7] Grachev P Yu, Ezhova E V and Sarapulov F N RU Patent 2 275 729 Winding of electric machine 2004.

[8] Grachev P Yu and Gorbachev E E RU Patent 2 526 835 Energy-efficient electrical machine 2014.

[9] Grachev P Yu, Gorbachev E E and Strizhakova E V RU Patent 2 602 551 Energy-efficiency electric machine with compact overhang parts 2016.

[10] Grachev P Yu, Bazarov A A and Tabachinskiy A S 2017 Compact Windings Design for Traction Induction Motors and SCIGs *International Conference on Industrial Engineering, Applications and Manufacturing (ICIEAM);* DOI: 10.1109/ICIEAM.2017.8076322.

[11] Grachev P Yu, Gorshkov R G and Makarov D V 2014 Application of windings with minimal overhang parts in electrotechnology *International scientific conference on topical problems of energy saving electrotechnology* pp 141-143.

[12] Bespalov V Y, Kobelev A S, Kruglikov O V and Makarov L N 2015 Constructing and mastering the production of energy-efficient asynchronous motors of mass-market series *Russian Electrical Engineering* vol 86 (4) pp 194-200.

[13] Grachev P Yu, Bazarov A A and Tabachinskiy A S 2017 Electrical and thermal processes of HEV induction machines taking into account stator winding form *IEEE Dynamics of Systems, Mechanisms and Machines (Dynamics).*

[14] Grachev P Yu, Gorbachev E E and Karsakov A Yu 2013 Efficient brushed machines with cost-saving armature winding *Efficiency improvement of brushed electromechanical converters application* pp 306-310.

[15] Inkin A I Electromagnetic fields and electric machines paratemers. Novosibirsk 2002 464 p.

[16] Chen Q, Kato S and Sawaya K 2008 Estimation of current distribution on multilayer printed circuit board by near-ld measurement *IEEE Transactions on Electromagnetic Compatibility* vol 50 (2) pp. 399-405.

[17] Petersen T H, Carpenter K H and May C M 2009 Comparison of experimental measurements of current distribution in a conductor with simulated results from the partial inductance method *IEEE Transactions on Electromagnetic Compatibility* vol 51(2) pp. 345-350.

[18] Gerasimenko T N, Ivanov V I, Polyakov P A and Popov Yu V 2011 Application of a conformal-mapping technique to a boundary-value problem of current distribution in plain conductors *Journal of Mathematical Sciences* vol 172(6) pp. 761-769.

[19] Grachev P Y, Strizhakova E V and Tabachinskiy A S 2017 Starter-Generator Design and Analysis of Start/Stop Processes for HEVs *International Conference on Industrial Engineering, Applications and Manufacturing (ICIEAM),* DOI: 10.1109/ICIEAM.2017.8076325.

[20] Kazakov Y B and Shvetsov N K 2015 Influence of pulse width modulated supply voltage waveform on energy efficiency and electromagnetic torque of induction motors *Acta Technica CSAV (Ceskoslovensk Akademie Ved)* vol 60 issue 3 pp 307-318.

[21] Vasilev B Yu and Kozyaruk A E 2013 Structure, composition, and control algorithms of high-efficiency electric drives of gas-compressor units *Russian Electrical Engineering* vol 84 issue 2 pp 94-102.

[22] Abakumov A M, Antropov V E and Randin D G 2017 Electrotechnical vibration isolation system with a magnetorheological damper *International Conference on Industrial Engineering, Applications and Manufacturing (ICIEAM),* DOI: 10.1109/ICIEAM.2017.8076160.

On synthesis of power regulator controllers of arc furnaces according to desired regulating characteristics

V A Ivanushkin[1] , D V Isakov[1], S E Mironov[2] and F N Sarapulov[3]

[1]Associate Professor, Ural Federal University (branch), Nizhniy Tagil, Russian Federation.
[2]Associate Professor, Ural Federal University, Ekaterinburg, Russian Federation.
[3]Professor, Ural Federal University, Ekaterinburg, Russian Federation.

E-mail: atps_nt@mail.ru,

Abstract. The paper considers a synthesis method of power regulator controllers of electric arc furnaces on the basis of unit typical non-linear functions; the main stages of synthesis of control devices; the mathematical model of the basic elementary unit of nonlinearities; the presented modular structure of the control unit of the power regulator of an arc furnace.

1. Introduction

With the use of programmable logic controllers (PLC) and modern variable-frequency electric AC drives in automatic power regulators of electric arc furnaces there have appeared new possibilities to improve quality of electric mode control and core technical and economic indicators of furnace operation.

It is known that the quality of the electric mode control system of the furnace is mainly defined by regulating characteristics (relations of free running electrode speed in the function of misalignment signal) of the electrode movement drive [1]. Implementation of desired regulating characteristics in the acting regulators was traditionally enforced within the capabilities of the electrode drive control system. Under this approach addressing the targeted task was narrowed down to applying negative nonlinear feedback in the drive coordinates (speed and voltage of the supply converter, etc.), as well as to including nonlinear functional converters in the direct transmission (mainly, in terms of the dead zone) [1]. Such approach inevitably resulted in necessity for designing special-purpose electrode drives that completed two tasks: controlling the drive coordinates as such and setting the object control law (relay, proportional, proportional-relay, or a more complex one).

At the present time during new detail engineering and upgrading operating mechanisms of power regulators of electric arc steelmaking furnaces using four-quadrant drives (for example, complete variable-frequency) it is rational to impose on the programmable logic controllers a task of formation of the desired regulating characteristics adapted to different melting periods. In this case the developer of the arc automatic regulator controller considers the complete controlled drive as an element that refers to the unalterable part of the controlled entity, thereby there is no longer any necessity to propose any needed changes in accordance with requirements to the structure and algorithms of the electric drive control system. Besides, in this particular case the developer of controllers brings into focus not the electric drive system coordinates control, but the implementation of the control algorithm of the definite process flow stage.

Under this approach there appears a task of choosing the representation and synthesis of different regulating characteristics (both symmetrical and asymmetrical) of the regulator controllers adopted to different melting process technological periods. In paper [2] there are stated the basic approaches to develop the power-efficient control algorithms of the electrical mode of the direct electric arc steelmaking furnaces taking into account the possible availability of modern PLC. The same article specifies the desired regulating characteristics of the automatic power regulators for different melting process periods containing piecewise-linear and piecewise-constant areas. Paper [3] suggests a method of analytical description of the similar non-linear functions of the regulators using function type sign (X) and abs (X) requiring the preliminary partition of the initial non-linear function to elementary standard components and their mathematical description.

Besides this, while the power regulators regulating characteristics being developed there arises a task of implementing non-specific non-linear functions for example when making an assessment and eliminating residual misalignment signal, as well as implementing smart regulation during furnace charge melt regime, the following non-linear functions are used: three-step relay characteristics along with the dead zone, sign selection, dead zone offset, etc [4,5,6,7].

2. The implementation of power regulators arc furnace

Practice shows that when using the mentioned method complex non-linear dependences typical for arc furnace, power regulators are possible to put into practice if using unit non-specific non-linear functions. The proposed method of synthesis provides the sequential performance of the following stages:

1. Requiring the desired characteristics of the controller in graphical form (for example, as represented by Figure 1) containing piecewise-linear and piecewise-constant areas with the desired cut coordinates (setup variables a_i, b_i).
2. Decomposition of the initial characteristics to the composition of simple components (symmetrical elementary typical non-linears).
3. Analytical description of every component using the models of the basic elementary static and dynamic non-linear unit functions (Table 1).
4. Formatting unit elementary characteristics to the real scale on both axis (it is necessary because the full level of the output signal of the actual components of the desired characteristics is not a single one and the interval of their action has definite values).
5. Structuring block-modular construction of the controllers (Figure 2) and algorithmic schemes of its module.).

Table 1 represents the mathematical models of the basic elementary static and dynamic unit non-linears that are used while there being developed the specified characteristics (Figure 1) of the electromechanical regulator control of the electric arc furnace with the integrating actuating mechanism of the variable speed. The block-modular construction of the controllers is shown in Figure 2. The block consists of two subblocks I and II generating a control signal under positive and negative error mismatch (on lifting and lowering the electrode respectively), subblock III of the misalignment sign indication, as well as subblock IV of the continuous monitoring of the residual misalignment and regulator performance rating.

Subblock I contains modules 1-7 that generate the control action on the electrode movement drive under positive misalignment signal (module (7) in Figure 2. characteristic (b) in Figure 1) in mode of reversible and nonreversible control. Upon that, the reversible control mode is implemented with the help of modules 1-4 (characteristic (a) in figure 1, components (1,2,3,4) in figure 2) and module 6 (Figure 2, characteristic (g) in Figure 1), given that the external signal $\alpha = 0$, $\beta = 1$. In nonreversible control mode (except for continuous short circuit fault) there is implemented non-type, mixed non-linear characteristics (characteristic (a) in figure 1, path – 0 → a4 → b4 → a3 → b3 → b2 → b1 → a1→ 0) with the help of the modules 1,2,3,4 and module 5 (characteristic (f) in Figure 1).

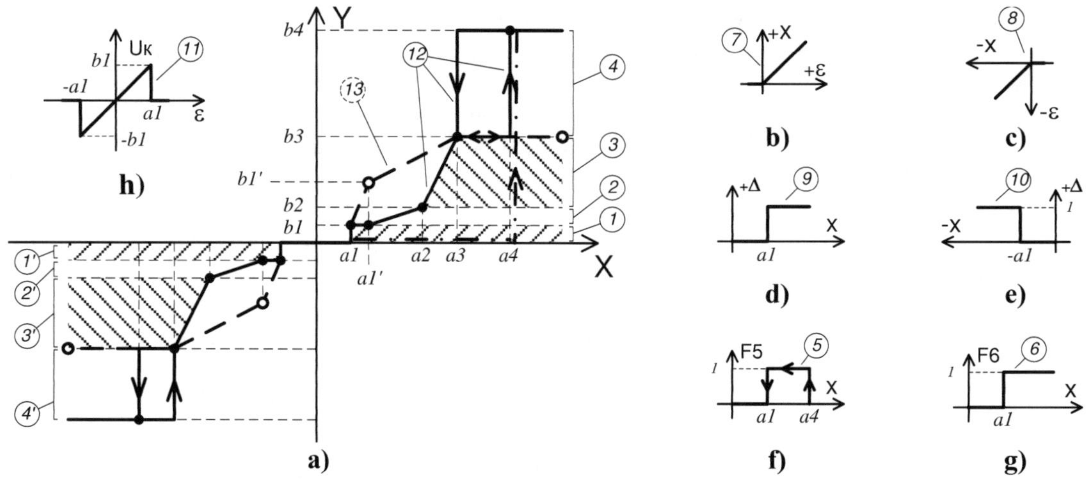

Figure 1. Statistic characteristics of the functional non-linear converter
(**a**) main view with elementary non-linear components (1-4 modules of its block-modular construction (figure 2)); (**b**, **c**) selection of positive and negative misalignment (7, 8 modules of its block-modular construction (figure 2)); (**d**, **e**) misalignment sign indication (9, 10 modules of its block-modular construction (figure 2)); (**f**, **g**) implementing smart regulation (5, 6 modules of its block-modular construction (figure 2)); (**h**) control of the residual misalignments (11 modules of its block-modular construction (figure 2)).

Table 1. Elementary unit non-linears

Name of non-linear	Mathematical description (model)
Three-step relay characteristics	$f(x) = \frac{1}{2}[sign(x-a) + sign(x+a)]$
Output limit on dead zone offset	$f(x) = \frac{1}{2}(\lvert x+b\rvert - \lvert x-b\rvert - \lvert x+a\rvert + \lvert x-a\rvert),\ b > a$
Three-step relay characteristics with hysteresis	$f(x) = \frac{1}{2}\left[sign\left(x - a\cdot sign\frac{dx}{dt}\right) + sign\left(x + b\cdot sign\left(x + b\cdot sign\frac{dx}{dt}\right)\right)\right]$ $b > a$
Selection of negative misalignment ε	$x(\varepsilon) = \frac{1}{2}(\varepsilon - \lvert\varepsilon\rvert)$
Selection of positive misalignment ε	$x(\varepsilon) = \frac{1}{2}(\varepsilon + \lvert\varepsilon\rvert)$
Three-step inverse relay characteristics	$f(x) = -\frac{1}{2}[sign(x-a) + sign(x+a)]$
Dead zone	$f(x) = \varepsilon - \frac{1}{2}(\lvert x+a\rvert - \lvert x-a\rvert)$
Output limit	$f(x) = \frac{1}{2}(\lvert x+a\rvert - \lvert x-a\rvert)$

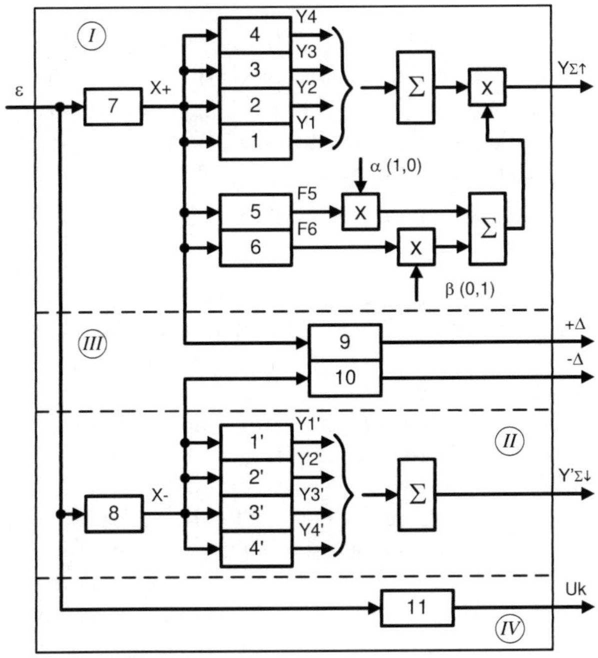

Figure 2. Block-modular construction of the controller.
I, II – subblocks of control signals formation under positive and negative misalignments;
III – subblock of sign misalignment indication;
IV – subblock of continuous misalignment monitoring and regulator performance rating.

The given mode setting condition is the presence of inverse values of the external signals α and β. With the help of module 5 under positive misalignment signal lowering there is provided an automatic bypass from the relay control to the proportional one.

Subblock II contains modules 8 (Figure 2, characteristic (c) in Figure 1) and 1′ - 4′ providing the formation of the desired controls while regulator processing the negative misalignment signals.

Subblock III contains modules 9 and 10 and generates the signals: +Δ – running of the regulated parameter over the upper bound of the dead zone (characteristic (d) in Figure 1); -Δ - running of the regulated parameter over the lower bound of the dead zone (characteristic (e) in Figure 1).

Subblock IV contains one module 11 implementing non-type non-linear static characteristics (characteristics (h) in Figure 1).

Algorithmic structures of all the subblocks modules are developed on the basis of mathematical models in Table 1. Table 2 shows unit non-liners that are used while subblocks (1-4) modules are being developed, as well as the coordinates of their setup variables and formatting indexes.

Table 2. Parameters and indexes of formatting unit non-linears

№ module	№№ non-linears	X-coordinate	Y-coordinate	Formatting index
1	*1*	*a1*	*b1*	*b1*
2	*2*	*a1, a2*	*b1, b2*	*b2 - b1 / a2 - a1*
3	*2*	*a2, a3*	*b2, b3*	*b3 – b2 / a3 – a2*
4	*3*	*a3, a4*	*b3, b4*	*b4 – b3*
5	*3*	*a1, a4*	*1*	*1*
6	*1*	*a1*	*1*	*1*
7	*4*	*0*	*0*	*1*
8	*5*	*0*	*0*	*1*
9	*1*	*a1*	*1*	*1*
10	*8*	*a1*	*1*	*1*
11	*7 + 8*	*a1*	*b1*	*b1*

IOP Conf. Series: Materials Science and Engineering **313** (2018) 012005 doi:10.1088/1757-899X/313/1/012005

3. Conclusion

It must be noted that the models given in Table 1 make it possible to design the desired regulating characteristics of the automatic regulator with integrating constant speed actuators under relay control.

The applicability of the proposed approach to the synthesis of power regulators controllers of electric arc furnaces on the basis of the desired regulating characteristics is determined by advisability of their implementation on the programmable logic controllers base.

References

[1] Pirozhnikov V E and Kablukovskii A F 1974 *Avtomatizatsiia kontrolia i upravleniia elektrostaleplavil'nymi ustanovkami* [Automation and control for electric arc furnaces plants] (Moscow: Metallurgiia – Press)

[2] Ivanushkin V A, Isakov D V, Sarapulov F N, Pozdeev S A and Rabek A A 2015 Formirovanie energo-effektivnykh algoritmov upravleniia elektricheskim rezhimom dugovykh stale-plavil'nykh pechei [Developing power-efficient control algorithms of the steelmaking arc furnaces electrical mode] *Industrial Power Engineering* **7** 32-35

[3] Ivanushkin V A, Isakov D V, Gamov A V and Sarapulov F N 2017 K realizatsii modelei upravliaiushchikh ustroistv s zadannymi regulirovochnymi kharakteristikami reguliatorov moshchnosti dugovykh elektropechei [On implementing models of the controllers with set regulating characteristics of the arc furnaces power regulators] *Effektivnoe i kachestvennoe snabzhenie i ispol'zovanie energii: sb. dokl. 6-oi mezhdunar. konf. v ramkakh spetsializir. foruma «Expo Build Russia»* Ekaterinburg 126-129

[4] Ivanushkin V A 1998 Reguliator moshchnosti dugovoi elektropechi [Power control electric arc furnace] *Certificate of authorship USSR* no. 1365375

[5] Ivanushkin V A 1988 Reguliator moshchnosti dugovoi elektropechi [Power control electric arc furnace] *Certificate of authorship USSR* no. 1424702

[6] Ivanushkin V A, Sarapulov F N, Isakov D V, Terekhov A Iu and Gamova T V 2012 Reguliator moshchnosti dugovoi staleplavil'noi pechi [Power control electric arc furnace] *Patent Russia* no. 2448165

[7] Ivanushkin V A, Sarapulov F N, Kozheurov V N and Isakōv D V 2008 Reguliator polozheniia elektroda dugovoi staleplavil'noi pechi [The position control electrode arc furnace] *Patent Russia* no. 2334926

IOP Conf. Series: Materials Science and Engineering **313** (2018) 012006 doi:10.1088/1757-899X/313/1/012006

Modern Reactive Power Generators

L. Chubraeva[1], **S. Timofeev**[2]

[1] Head of the laboratory of electric power industry, Institute of Electrophysics and Power-engineering of RAS, Saint-Petersburg, Russia
[2] Research scientist, Institute of Electrophysics and Power-engineering of RAS, Saint-Petersburg, Russia

E-mail: lidiach@mail.ru

Abstract. The paper reviews main stages of development of reactive power generators, describes the 1-st and 2-nd generation of synchronous condensers with conventional cooling systems and a new generation – superconductive synchronous condensers. Asynchronous non-salient pole condensers expand the class of rotating compensating devices. Comparison of dynamic performance of conventional synchronous condensers, cryogenic condensers and SVC is presented. The variant of a model 5 MVA HTSC synchronous condenser intended for wind power plants is described.

1. Introduction

Reactive power is the power of AC circuits utilized for supporting magnetic fields in inductive loads (transformers, electrical motors, inductive stoves, etc.) or electric fields for capacitors, cables, etc. The specific feature of reactive power is the fact that it cannot be transformed into another form of energy.

Reactive power compensation is defined as reactive power management with the aim of improving the performance of AC power [1, 2]. Reactive power compensation is considered in two aspects: load compensation and voltage support. In load support the objectives are to increase the power factor, to balance the load and to eliminate current harmonics from nonlinear industrial loads [1]. Voltage support reduces voltage fluctuations at a certain terminal of transmission line and improves the stability of AC system by increasing the maximum active power that can be transmitted It helps to attain a flat voltage profile at all levels of power transmission, increases transmission efficiency, controls steady-state and temporary over-voltages and helps to avoid catastrophic black-outs [1].

Synchronous condensers allow to increase the throughout transmission capacity of individual transmission lines, intersystem or interstate long distance transmission lines. They are responsible for correction of electrical power flow along the circuits with different voltage values in multi-contour electrical grids to obtain positive technical or commercial effect [2].

As reactive power compensation is an effective way to improve the electric power network, there is an urgent need for controlled reactive power generation. Reactive power generator is one of the devices, responsible to solve the above mentioned tasks.

Previously reactive power generators and synchronous condensers were synonyms. They represented a multi-pole synchronous machine without mechanical load on the shaft, operating with variable excitation and armature currents in accordance with U-curves [3].

In Russia there exists an excess of volume of reactive power in transmission networks. In 500 kV lines the level of compensation equals approximately 45 %. In the 330 and 200 kV transmission lines reactive power is practically not compensated [2].

IOP Conf. Series: Materials Science and Engineering **313** (2018) 012006 doi:10.1088/1757-899X/313/1/012006

2. First generation of conventional synchronous condensers

The development of synchronous condensers started in 1911 by the General Electric (USA). The first alternator rated at 10 MVA was manufactured in 1919 [4]. It was an air-cooled machine, ordered by Ontario Hydro.

Till the end of 70-s of the previous century synchronous condensers as reactive power generators dominated in electric grids, they were manufactured by numerous electrical machine-building companies with the largest one produced and tested in Sweden [5]. It had the rating 345 MVA and was a fully water-cooled machine, intended for Dumont station.

At 70-s synchronous condensers started to be replaced by electronic devices: static reactive power compensators (SVC – static VAR compensator, STATCOM – static shunt compensator), based on application of power electronics, capacitor banks and reactors. They possessed certain advantages, including fast response ability. Later flexible AS transmission systems (FACTS) were introduced.

Synchronous condensers stayed to operate, working out their resource. There appeared publications, justifying the advantages of rotating electrical devices [6]. Nevertheless the 1-st generation of synchronous condensers manufacturing with mainly air- and hydrogen cooling systems faded out about 1980-s.

Around 15 years ago a revival of rotating synchronous condensers could be witnessed in many countries. The change in energy mix is leading to replacement of conventional energy production with synchronous generators by low inertia and low thermal time constant power park modules [7]. These park modules are represented by wind farms and solar power plants.

As a result the second generation of conventional synchronous condensers started to develop alongside with one more type of reactive power generators – asynchronous condensers.

3. Second generation of conventional synchronous condensers

The era of the 2-nd generation of synchronous condensers started beginning from 2000 [4]. Main manufacturers are ABB, GE, Siemens. The unit rating exceeds 400 MVA. There practical application is associated with wind- and solar power installations and high-voltage DC transmission lines (HVDC). Their main advantages are evaluated as follows:

- increased short-circuit MVA strength,
- response to lower system fluctuations,
- improved high-response excitation systems,
- voltage and frequency ride-through capability,
- overload capability,
- high short and mid-term overload capability,
- improved reliability and increased service life, low maintenance.

Figure. 1 represents the design of synchronous condenser, manufactured by GE. It is a salient four pole alternator.

There are developed different methods of synchronizing with the network: with pony motors or with reactors in the armature circuit, depending on the demands of the consumer.

As an example the plan of installation of synchronous condensers in Denmark is shown in figure.2 [8].

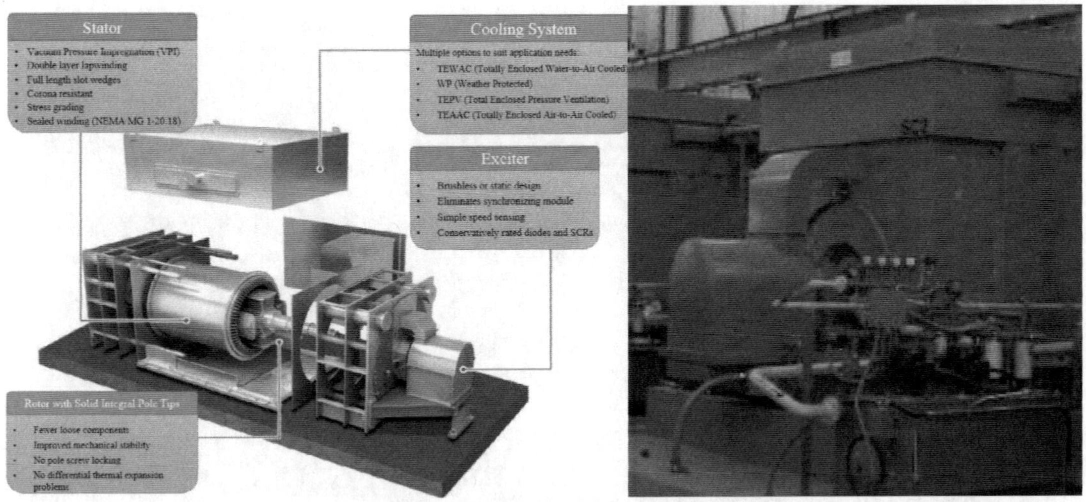

Figure 1. Illustrations referring to GE 2-nd generation synchronous condenser

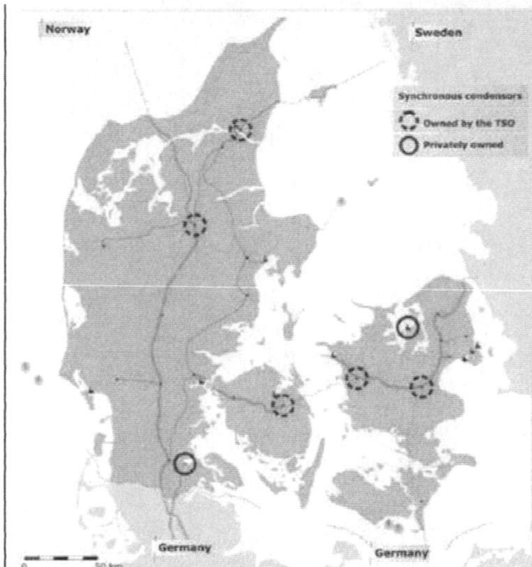

Station	Area	PCC kV	Sn MVA	Year
Vr. Hassing	DK1	150	100	1965/2013
Tjele	DK1	150	160	1976/2012
Bjæverskov	DK2	400	270	2013
Herslev	DK2	400	200	2014
Fraugde	DK1	400	200	2014
Kyndbyværket	DK2	132	330	1976/199x
Enstedværket	DK1	150	880	1979/2013

Figure 2. Application of 2-nd generation of synchronous condenser in Denmark

4. Asynchronous condensers

After manufacturing and successful exploitation of a number of asynchronous turbogenerators there was developed and manufactured by OJSC «Power Machines» a non-salient pole air-cooled asynchronous condenser rated at 100 MVA (figure. 3) [2,]. Two alternators were installed in a Beskudnikovo 500 kV substation (Russia, Moscow region). Main parameters: voltage - 20 kV, frequency of rotation - 1500 rpm, total mass – 250 t. The main difference as compared to synchronous condensers lies in the fact that the rotor contains two excitation windings. The second one is in the q-axis. Magnetomotive force of the second winding is less than the one of the main winding. Two windings are operating with two excitation systems.

IOP Conf. Series: Materials Science and Engineering **313** (2018) 012006 doi:10.1088/1757-899X/313/1/012006

Figure 3. Asynchronous condenser rated at 100 MVA

Advantages of asynchronous condensers with non-salient pole rotor are as follows [2]:

- Wide range of reactive power regulation from 100% to -100%.
- Higher response of reactive power or voltage regulation due to possibility of current reverse in excitation windings.
- Better damping of rotor oscillations during disturbances in the network.
- High survivability due to presence of two excitation windings.

It should be nevertheless noted that a non-salient pole machine with two excitation windings and two excitation systems is relatively expensive.

5. Superconductive synchronous condensers

In 80-s - 90-s there were manufactured and tested several superconductive generators which showed they may operate as synchronous condensers with the rated reactive power generation and consumption due to better U-curve and more favorable relation of magnetomotive forces of excitation and armature windings.

It allowed to exclude the limitations in the under-excitation mode because of better summary magnetic fields in the armature end-zone.

The 20 MVA, 3000 rpm superconducting generator was manufactured in the USSR in 1979 (figure. 4, a).The excitation winding was wound of Nb-Ti low-temperature superconductor (LTSC).

From 1982 it was operating as a synchronous condenser in the network “Lenenergo” and allowed to reveal advantages as compared to SVC. The peculiar feature of this alternator was the fact that the armature was cooled by an insulating liquid and the armature cooling system was equipped with a flux-pump. The same liquid was a working medium of the flux-pump. The armature losses were used for water heating [9].

In Japan there were produced an LTSC synchronous condensers rated at 30 MVA and 70 MVA LTSC generator, which operated in Kansai Electric Power Company grid in Osaka synchronous condenser (figure. 4, b) [10].

In 1986 high-temperature superconductivity (HTSC) was discovered and stopped the works with LTSC alternators due to the main advantage of HTSC – higher operating critical temperature.

Basing on obtained practical experience with LTSC synchronous condensers the USA started development and manufacturing of a new generation of synchronous condensers – dynamic HTSC synchronous condensers.

Dynamic synchronous condenser design is based on HTSC field winding, which requires only a small change of current in under- over-excitation modes due to low synchronous reactance.

There was developed a trailized 8 MVA HTSC condenser (figure. 5.), which was installed in 2004 on the Tennesy Valley Authority (TVA) grid and showed positive results of operation. TVA ordered several HTSC condensers of higher rating [11, 12].

a b

Figure 4. Helium-cooled 20 MVA alternator (Russia) and 70 MVA alternator (Japan).

To evaluate a possibility of development of a high-response synchronous condenser we have developed a project of high-voltage synchronous condenser rated at 200 MVA, 220 kV with increased excitation current, decreased reactance and original electromagnetic screen on the rotor (figure. 6). The main demands are presented below. Comparison with other types of synchronous condensers is given in Table 1.

The preliminary evaluation of the variant for 550 kV revealed a lot of technical problems. One of them showed it is practically impossible to set the high-voltage terminals on the condenser outer diameter of the frame because of relatively small machine size.

The machine was developed according to the following data:

- Electric voltage – 220-500 kV.
- Armature overload – double for 60 s.
- Range of reactive power variation Q = ± (160 - 200) MVA.
- Time of attaining of rating from Qc = 0 to Qc = ±(160 - 200) MVA - 0.4 – 0.6 s.
- Synchronous inductance – xd = xq - (0.3 - 0.6) p.u.

IOP Conf. Series: Materials Science and Engineering **313** (2018) 012006 doi:10.1088/1757-899X/313/1/012006

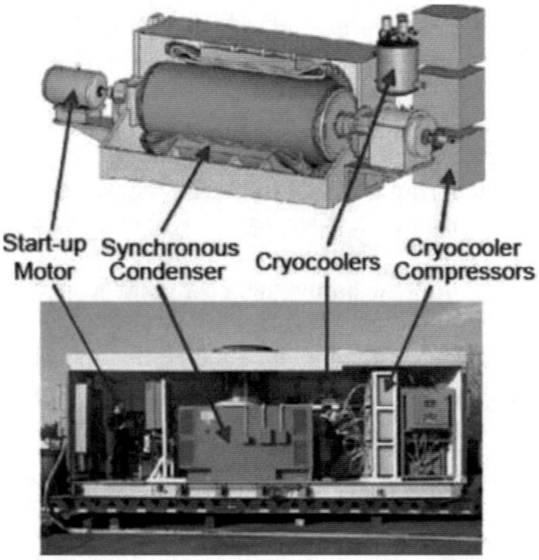

Figure 5. Trailized 8 MVA SuperVAR HTSC synchronous condenser

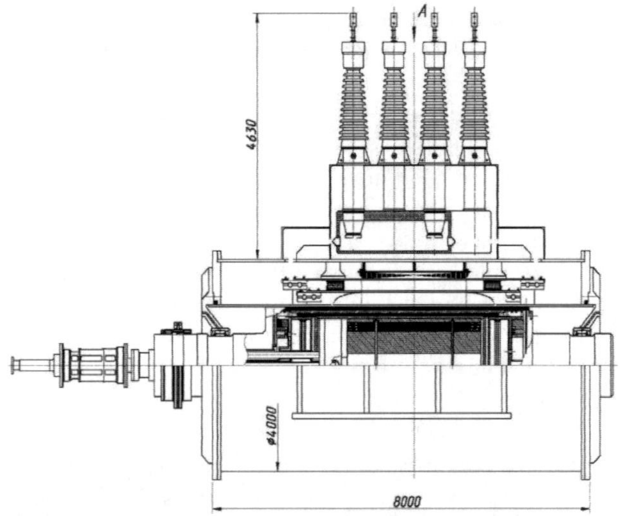

Figure 6. Longitudinal cross-section of 200 MVA HTSC synchronous condensers

Table 1. Comparison of conventional and cryogenic synchronous condensers

Parameters	Value		
	Cooling system		
	Hydrogen	Water	Cryogenic
Rated power, MVA	160	345	200
Linear voltage, kV	15.75	20	220
Current, A	5860	10100	525
Rotor speed of rotation, rev/min	750	750	750
Frequency, Hz	50	50	50
Number of phases	3	3	3
Stator diameter, mm	2500	4000	3270
Rotor diameter	2440	3800	2200
Active length, mm	3200	3600	2200
Stator yoke height, mm	375	-	120
Housing diameter, mm	5200	6500	4000
Synchronous reactance, p.u.	2.05	1.9	1.0
Transient synchronous reactance, p.u.	0.45	0.502	0.58
Subtransient synchronous reactance p.u.	0.21	0.245	0.37

The machine was developed according with the following data:

- Electric voltage – 220-500 kV.
- Armature overload – double for 60 s.
- Range of reactive power variation Q = ± (160-200) MVA.
- Time of attaining of rating from Qc = 0 to Qc = ±(160-200) MVA - 0.4 – 0.6 s.
- Synchronous inductance – xd = xq = (0.3 - 0.6) p.u.

Main advantages of HTSC synchronous condensers are considered to be as follows:

- wide range of reactive power regulation,
- possibility of twofold overload,
- absence of thermal stresses in the cryogenic windings due to stable operating temperature,
- improved reliability,
- decreased losses, comparable with static compensating devices,
- lower mass and size indexes,
- highly sinusoidal output voltage,
- decreased synchronous reactance value and increased limits for reactive power regulation in the under-excitation mode,
- possibility of fast response to transient voltage variation,
- possibility of unit rating increase,
- possibility of operating at zero field current at xd=1 p.u,,
- feasibility of high output voltage (up to 200 kV and more).

Comparison of U-curves and losses is given in figure. 7.

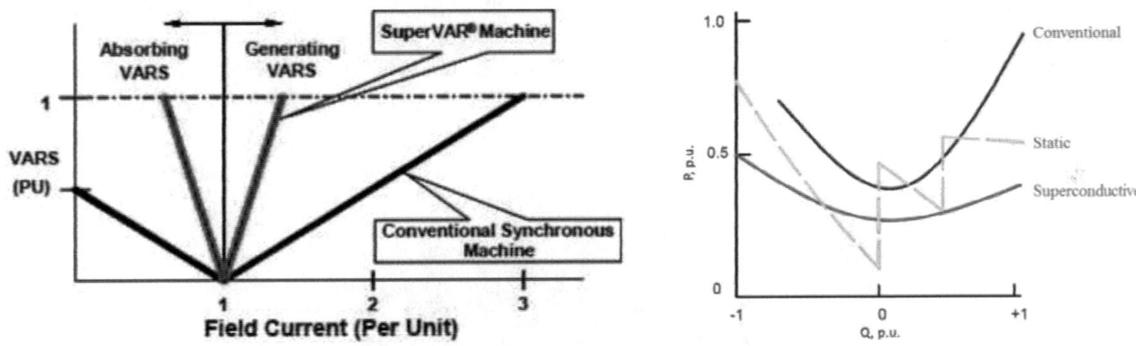

Figure 7. U- curves and loss variation for different types of compensating devices.

6. Comparison of dynamic performance of compensating devices

In [1] a comparison of dynamic operation of different types of reactive power compensating devices is performed. Simulaton of different fault types show the following.

In case of a single phase to ground fault the SVC injects reactive power after a delay caused by the delay in the controller, but after a delay it injects more reactive power due to faster voltage control. The synchronous condenser reacts instantly due to the change in the thermal voltage. Superconductive condenser injects more reactive power at the instant of the fault as compared to conventional one due to lower synchronous reactance. But at the fault clearance both types of condensers act similar.

During a two-phase to ground fault the positive-sequence voltage reaches the rated value faster with both types of synchronous condensers. It is because SVC injects less reactive power during and after the fault until the line is connected back.

In case of three phase to ground fault the positive-sequence voltage drop is still larger with the SVC. The reactive power injected by both synchronous condenser types is still higher. The results are given in Tables 2 and 3.

Table 2. Minimum positive sequence voltage during the fault at 36-kV bus

Fault Type	1ph	2ph	3ph
Min. Ter. V (p.u.)			
With SVC	0.89	0.60	0.30*
With Conv. S.C.	0.89	0.67	0.42
With SuperVAR	0.89	0.70	0.45

IOP Conf. Series: Materials Science and Engineering **313** (2018) 012006 doi:10.1088/1757-899X/313/1/012006

Table 3. Maximum injected reactive power by synchronous condensers and SVC during the fault

Fault Type	1ph	2ph	3ph
Injected Q (p.u.)			
SVC	0.67	0.48	0.10*
Conv. S.C.	0.54	0.69	0.74
SuperVAR	0.54	0.93	1.07

During severe faults synchronous condensers perform better and bring the terminal voltage to rated value faster.

A very interesting experimental investigation was carried out at Vyborg substation (Leningrad region) with parallel operation of conventional synchronous condenser and STATCOM [13].

7. Model 5MVA HTSC synchronous condenser

There is now under development by our team a version of HTSC 5 MVA, 750 rpm, 10.5 kV prototype SC for wind power plants [14]. It should be specially noted that all the variants have LN2 cooling system. On the first stage there was discussed a variant with a slotless armature as well. But later on it was rejected because it came out to be more expensive, than the 1-st variant of SC, described later.

Below are presented the results of development of two variants of SC. The 1-st one contains a 2G HTSC 8 pole field winding placed in the cold zone (figure. 8).

Figure 8. A variant with 2G HTSC 8-pole field winding and conventional armature winding

The rotor represents an 8-pole design with salient poles and field winding with concentric coils. The magnetic core is positioned inside the cryostat. The field winding coils are produced of 2G HTSC tape. The electromagnetic screen and the damper winding are provided in the rotor design to ensure reliable operation. The vacuum cavities reduce heat transfer to the cold zone and decrease the flow of cooling agent during the SC operation. When developing the rotor design during the cool down process the coefficients of linear shortening of the elements were taken into account. The synchronous condenser stator is a conventional one to reduce, as mentioned above, the cost of the new machine.

The second variant is based on the design with HTSC armature. The armature winding contains 12 saddle-shape coils of 2G HTSC, positioned in a single cryostat with amorphous alloy core. The coils are laid out in special slots formed of amorphous alloy tape and are fixed from the rotor side by fiber glass cylinder (figure. 9). The outer and side surfaces of the cryostat are made of stainless steel

The armature current value equals 262 A, therefore the coils are being wound of 4 HTSC tapes in parallel. We do not consider it reasonable to transpose tapes along the coil length and are making the transposition when connecting the coils with each other. As a result the circuit branch for circulating currents increases. Such a method of transposing saves space and simplifies the process of winding manufacturing. The method was successfully implied during the 20 MVA LTSC turbogenerator armature manufacturing.

Main data of calculations for both variants of SC is presented below in Table 4.

Table 4. Comparison of two variants of HTSC synchronous condensers

Parameters of HTSC synchronous condenser	Value	HTSC rotor	HTSC stator
Rated total power	MVA	5.0	5.0
Rated linear voltage	kV	10.5	10.5
Rated speed of rotation	rpm	750	750
Rated phase voltage	кV	6.3	6.3
Rated current	A	262	262
Rated magnetic induction	T	0.82	0.82
Inner armature diameter	mm	1588	1777
Active armature length	mm	811	907
Number of armature winding bars/coils	unit	144	12
Armature current density	A/mm2	4.16	65
Air gap	mm	5	7
Synchronous reactance along longitudinal axis	p.u.	1.198	0.826
Synchronous reactance along quadrature axis	p.u.	0.671	0.587

8. Conclusions

Reactive power generators are represented now by both synchronous and asynchronous electrical machines. The renewed interest to rotating reactive power compensating devices is associated mainly with the development of wind- and solar power stations and high-voltage DC transmission lines. As a result the 2-nd generation of conventional synchronous condensers is developing.

The investigations of superconductive condensers are continued. In many aspects this alternators may compete successfully with static compensating devices and with conventional synchronous condensers as well. To win in this competition there is an urgent need for improved types of HTSC wires and cables, intended for high currents (several kA).

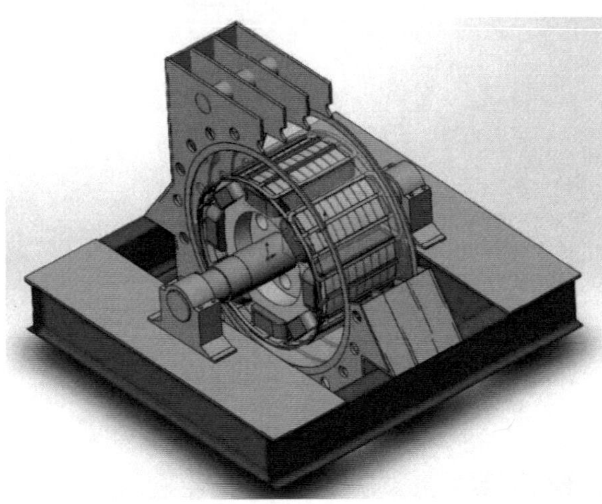

Figure 9. A variant with conventional 8-pole field winding and 2G HTSC armature winding

References

[1] Teleke S, Abdulahovic T, Thiringer T, and J Svensson J 2008 *IEEE Trans. on Power Delivery* **23** 1606-12
[2] Pinchuk N D, Antonyuk O V, Kadi-Ogly I A, Sidelnikov A V, Ioganзen V I, Dubrovin Yu N, Kartashova T N, Kiilo O L 2009 *Power Technology and Engineering* **43** 60-63
[3] Pelissier R 1982 *Les Reseaux d'Energie Electrique* (Moscow: High School) pp 85-116.
[4] D'Aguila R 2014 *Minnesota Power System Conference*, (Minnesota: Brooklyn Center)
[5] Landhuh H A and Nordberg B 1971 *IEEE Trans. On PAS* **90** 2765-77.
[6] Guillard J M et al. 1990 *CIGRE Report* 11-102 6 p
[7] Glaninger-Katschnig A 2013 *e & i Elektrotechnik und Informationstechnik* **130** 28-32
[8] Abidgaord H and Qin N 2015 *Ice-Breaking Projects in Power Conf.* (Ilmenau: Technische Universität Ilmenau)
[9] Chubraeva L I 1995 Synchronous condenser mode of a cryoalternator *Power Technology Conf.* (Stockholm) p 6
[10] Grant P M 1997 *Applied Superconductivity, IEEE Transactions on* **7** 112-133
[11] Kalsi S, Madura D and Ross M 2005 *IEEE/PES Transmission and Distribution Conf. & Exh. Asia and Pacific* (Dalian)
[12] Kalsi S, Madura D and Ingram M 2015 *IEEE Trans. on Applied Superconductivity* **15** 2146-49.
[13] Zamzanov Eu Ju et al. 2011 *Izvestija of RI on DC Current. Scientific Collection* **65** 19-26
[14] Chubraeva L I, Timofeev S S 2017 *Materials Science Forum* (in print).

Contactless system of excitation current measurement in the windings with high inductance

L. Chubraeva[1], E. Evseev[2], S. Timofeev[3]

[1] Head of the laboratory of electric power industry, Institute of Electrophysics and Power-engineering of RAS, Saint-Petersburg, Russia
[2] Research scientist, Institute of Electrophysics and Power-engineering of RAS, Saint-Petersburg, Russia
[3] Research scientist, Institute of Electrophysics and Power-engineering of RAS, Saint-Petersburg, Russia

E-mail: lidiach@mail.ru

Abstract. The results of development, manufacturing and testing of a special contactless maintenance-free excitation current measurement system intended for the windings with high inductance, typical for superconductive alternators, are presented. The system was assembled on the brushless exciter is intended for 1 MVA wind-power generator with the winding, manufactured of high-temperature superconductors (HTSC). The alternator with brushless exciter were manufactured and successfully tested.

1. Introduction

One of the specific features of alternators with superconducting excitation windings is high inductance of the windings. It refers to all types of alternators: generators, motors, synchronous condensers [1]. There is no major difference in low- or high-temperature superconductors in this aspect. The control of the excitation current value is important because of possible quench of the winding with transition from a superconductive to normal state. It results in appearance of high electric resistance and high voltage on the excitation winding terminals. Not to destroy the winding special protection system is envisioned. It gets the signals from the current and temperature measuring systems.

The current measuring system was developed for 1 MVA wind-power generator with HTSC excitation winding and brushless exciter. The alternator comprises a multi-pole winding. The coils are manufactured of HTSC tape of the 2-nd generation with relatively low current value and as a result high inductance value. The rotor has cryogenic cooling. The armature is of a conventional type. It was decided to make a brushless exciter operating with conventional air-cooling. It contains the main exciter and sub-exciter and is independent of external current supply [2]. Electrical connection of exciter with the excitation winding, i.e. copper wires with HTSC, is performed within the rotating shaft.

The whole complex was manufactured and successfully tested within the SC «Rosatom» Program «Superconductive Industry» [3].

The main aim of the work is the development of contactless maintenance-free system of excitation current measurements. It is to be responsible for the following: possibility of measurements in the current operating range; independence of the accuracy of measurements from the rotor frequency of rotation; insensitivity to external thermal, electromagnetic or mechanical influences; repeatability in case of serial production with application of a maximum possible amount of standard elements.

2. Overview of applied measurement system

Analysis of existing systems of excitation current measurement in turbogenerators with both independent and brushless excitation systems revealed several drawbacks, making them not applicable for superconductive wind-power alternator [4-8]. They either comprise subjected to wear rotating elements or are influenced by numerous factors mispresenting the final result. Moreover in many cases they have negative influence on the safety of operation. The latter is very important for the wind-power installations.

Specific feature of the developed superconductive alternator with brushless exciter is as mentioned above electrical connection of copper and HTSC current leads is performed inside the shaft of the device. Therefore even very attractive variants of patents [8] cannot be applied.

Modern measuring systems are mainly based on application of Hall sensors. The latter posses the following advantages: galvanic isolation of the prime and secondary (measuring) circuits, high sensibility, wide temperature range, etc. Measuring systems, based on Hall sensors, provide high reliability, very low total error and allow to decrease the volume of the device [9, 10].

3. Main features of the developed system

The wind power alternator operates with a brushless exciter. The measuring system is one of the elements of the complex.

During the system development process several principal conditions, obligatory for performance, were accepted: rejection of application of units subjected to wear, application of sensors, not possessing their own inductance, excluding or decreasing of units with residual magnetization, easiness of setting, calibration and exploitation.

Calculation and manufacturing of the system started from the rotating coil without magnetic core (figure. 1).

The coil consists of non-magnetic disc 1, contact bolts 2, two counter wound windings 3. The coil is fixed on the brushless exciter shaft 4 with current leads 5 within it, connecting the exciter with HTSC excitation winding.

The ELCUT calculating program, applied to the design, showed the magnetic induction value equals several mT within a distance of 10 mm from the coil. The value is absolutely adequate for application of modern magnetic field sensors.

To choose the element with proper magnetic sensibility the experience of serial industrial devices application, based on Hall sensors, magnetoresistances, magnetodiodes and magnetotransistors [11]. The choice was made in favor of the element, based on the design of magnetoelectronic reproducing head. The principal peculiarity of these meters is based on linearity of Hall sensor electromotive force characteristics up to the frequency of several MHz. The sensor design represents a Hall pick-up 4, positioned in additional air-gap 3 of ferrite circular core 2 (figure. 2, a).

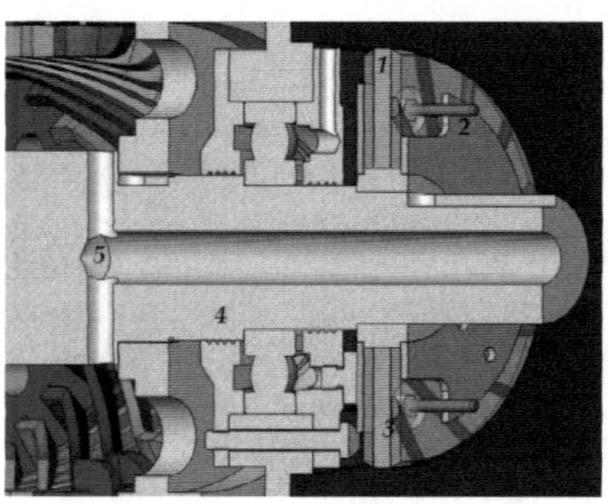

Figure 1. 3D cross-section of the unit, connecting brushless exciter with wind-power alternator HTSC winding.

IOP Conf. Series: Materials Science and Engineering **313** (2018) 012007 doi:10.1088/1757-899X/313/1/012007

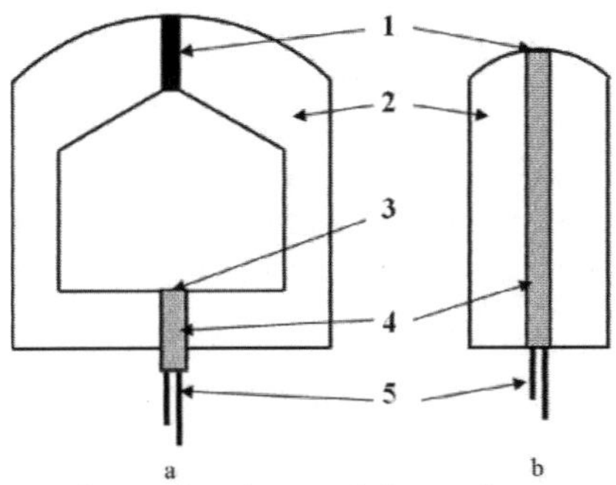

Figure 2. Principal scheme of magnetoelectronic reproducing head with Hall pick-ups

The main air-gap 1 faces the source of measured magnetic field.

The industrial magnetoelectronic heads possess high sensibility and a wide frequency range.

Therefore the sensor was manufactured on the base of a serial item, that was initially close to necessary frequency and magnetic field values (45-135 Hz at 300-900 rpm, 9-turn coil and magnetic field value 1.4 mT on a distance 5 mm from the sensor to the coil.

The chosen device was CSLA1CH (firm "Honeywell"), possessing the following parameters: peak voltage – 150 A, sensibility – m 19.6 mV/turn, feeding voltage equals 12 V.

4. Manufacturing and testing

The device [12] was subjected to revision with respect to the processing of the device frame and addition of operating magnetic gap, equal to 4 mm (figure. 3)

To obtain the characteristics of the improved device several tests with determination of the sensor sensibility and possibility of screening from undesirable magnetic fields influence (figure. 4) were carried out. In the first test the distance between the pick-up and the coil was 2 mm with feeding voltage 12 V. In the second test the influence of external magnetic field on the data of non-screened and magnetically screened device was measured.

The screen represents a steel tube with oval cross-section.

Figure 3. Sensor CSLA1CH prio to and after modification.

Experimental results allowed to conclude the sensor is workable at external magnetic fields on the screen surface up to 0.1 T.

As far as during the alternator operation it is necessary for the signal from the sensor to be close to sinusoidal one with the frequency, equal to the frequency of the alternator and the exciter rotation, a special high-speed microcontroller is applied for data processing. The coordination of the sensor output voltage level with the microcontroller input was performed.

It demanded to calculate and manufacture a low frequency first order filter based on operational amplifier of LM258N type with the cutoff frequency 10 kHz and coefficient of amplification equal to 16.

The process of assembly of the unit with microcontroller is presented in figure. 5.

The microcontroller type is «pic16f628a», the data processing is performed by specially developed algorithm.

The algorithm excludes mispresentation of measuring system readings because of sensor signal deviation with respect to zero. The microcontroller measures amplitudes of two adjacent signal surges and integrates the average value during one turn – it corresponds to 18 measurements. The presence of low frequency filter excludes high frequency disturbances in the scheme of level coherence.

Temperature drift and external magnetic fields have much lower frequency as compared to the frequency of measurements and are ignored by the microcontroller. As a result high accuracy of measurements is obtained.

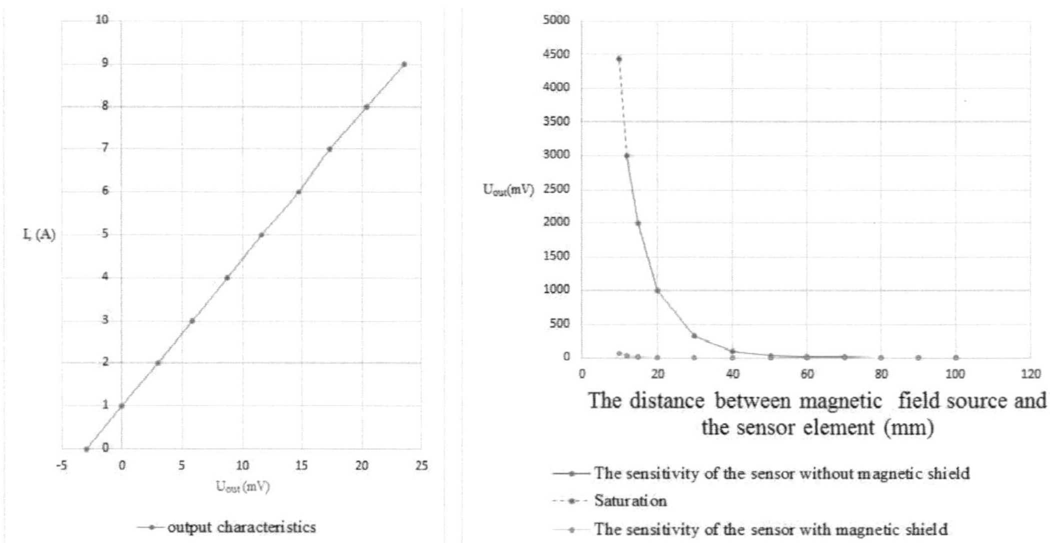

Figure 4. Sensor CSLA1CH experimental data: a)– graph of the sensor during tests, b) – curves, characterizing sensibility to external magnetic field.

Figure 5. Assembly of the microcontroller unit.

The results allowed to conclude that the sensor is efficient under the operating conditions of the alternator and brushless exciter. The device sensibility equalled 3 mV/A.

The scheme of final design is presented in figure. 6, a: 1- magnetic screen, 2 – sensor within the screen. The screen with the sensor is fixed to the exciter frame 3 above the rotating coil 4.

Electronic part with the sensor in magnetic screen was impregnated with epoxy. The final view of the contactless current measuring system, fixed on the frame of the brushless exciter is presented in Fig. 6, b.

IOP Conf. Series: Materials Science and Engineering **313** (2018) 012007 doi:10.1088/1757-899X/313/1/012007

After the system assembly the comprehensive testing was carried out current 60 A and maximum current 100 A. The experiments were carried out on a special test-bed targeted for the full scale investigations of the brushless exciter and contactless measuring system,

The results, referring to the measuring system, are shown in figure. 7. The data was obtained by a four channel digital oscilloscope.

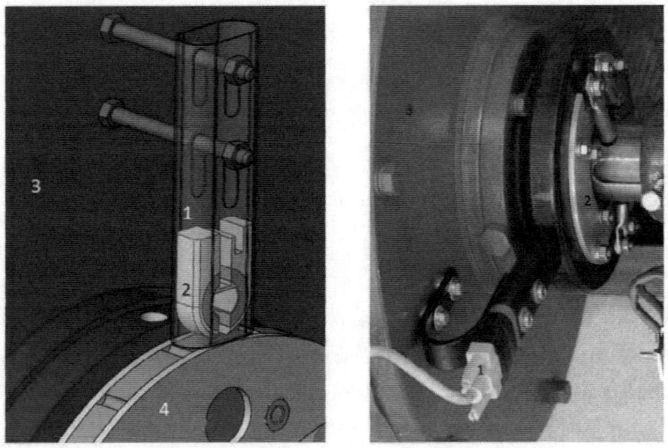

Figure 6. Measuring unit, assembled on brushless exciter: a) – schematic view, b) – view of the manufactured device.

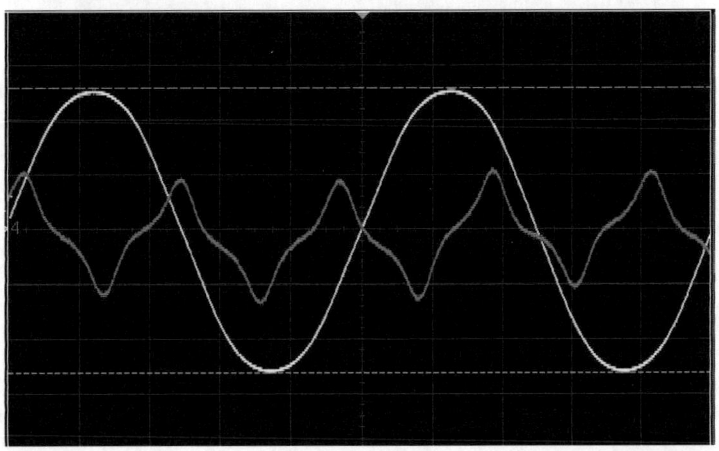

Figure 7. Oscillograms, obtained during the contactless measuring system tests

The pink curve represents the signal from the sensor. The curve has specific "steps" caused by relatively big coil winding step. A small deviation of about 250 mV may be witnessed as well due to the influence of the main bars, rotating with the exciter shaft (on the test-bed the exciter was tested individually, without the main alternator).

To present the signal form relative to the position of the shaft the first channel of the oscilloscope was switched to the phase of the sub-exciter AC winding, positioned on the main shaft (yellow curve). Four periods of the sinusoid correspond to one turn of the system.

The signal from the sensor corresponds to the calculated value and after its processing the true excitation current value may be obtained and in case of kIc excess (Ic – critical current of superconductor, k - safety factor) special measures of superconductive excitation winding protection from transition to normal (non-superconducting) state enter into force.

5. Conclusion

The results show a principal possibility to increase substantially the accuracy of contactless systems of electric current measurements. Moreover the systems may be maintenance free without periodical technical servicing. They may be applied for a variety of devices both conventional and superconductive ones.

References

[1] Danilevich Ja B, Chubraeva L I 1994 *New Generator Designs and Problems of Their Development* (Saint-Petersburg: Nauka) p 224
[2] Devyatova T A 2017 *Nuclear Strategy XXI* **128** 8-10
[3] Chubraeva L I, Timofeev S S 2016 *Materials Science Forum* **856** 38-43
[4] Charkow Kildischew W S, Kowalkow G A, Masnikin N S, Misjulin A W 1975 Contactless excitation current measurement of brushless generator *Patent DE 19712108671*
[5] Polyakov V I 2001 Measurement of rotor current in generators with brushless excitation *Electrical station* **2** 59-61
[6] Tsvetkov B.A. et al. 1988 Method of determination of short-circuits in rotor winding of synchronous electrical machine *Cop. Sert. 1436649 (USSR)* 41.
[7] Polyakov V I 1994 Method of rotor current measurement in a generator with brushless exciter *Patent 2011203*
[8] Popov I N et al. 2008 Method of measuring of rotor current in synchronous generator with brushless excitation *Patent № 2327273*
[9] Bolotin O.A. et al. 2016 Modern sensors for current and voltage measurement *ISUP* **1(61)** 5
[10] Gilardy M 2013 New horizons for sensors based on Hall effect *Power electronics* **3** 12-15
[11] Semjenov A L 2012 *Magnetic Materials for Micro- and Nanoelectronics* (Irkutsk: ISU) p 147
[12] Chubraeva L I et al. 2017 Way and generator rotor current measuring device with the brushless excitation *Patent 2623696 C1*

IOP Conf. Series: Materials Science and Engineering **313** (2018) 012008 doi:10.1088/1757-899X/313/1/012008

Power supplies for dual-frequency induction melting of metals

V I Lusgin[1,2], A S Koptyakov[1], A U Petrov[2], K A Zinovev[1] and D A Kamaev[1,2]

[1] High Voltage Technique Department, Ural Federal University, Ekaterinburg, Russian Federation
[2] RELTEC Ltd, Ekaterinburg, Russian Federation

E-mail: a.s.koptyakov@ya.ru

Abstract. The article discusses the benefits of multi frequency induction melting in the production of synthetic cast iron, structural (electric circuit) principles of dual frequency Power supplies of melting systems. The ways of electric power regulation of low frequency and high frequency components of the current in the inductor sections of furnace are demonstrated, namely power rescheduling at the metal melting stage, alloying stage and decarburizing of synthetic cast iron.

1. Introduction

Induction melting of metals in strong electromagnetic fields is widely used to obtain complex high homogenized alloys of ferrous and non-ferrous metals in modern foundry production of parts with high mechanical and physical characteristics. Undoubted advantages of medium frequency induction furnaces are: low energy consumption per ton of produced metal; batch regime of periodic melting (without "heel"); the high intensity of melting at high specific power of the electric power supplied to the metal; high performance of a metal production with a relatively low cost of induction furnace equipment.

Improving the performance of induction furnaces is provided by the increase of specific power at a higher current frequency in the furnace inductor [1]. However, both the power action of high frequency electromagnetic field to the molten metal, and the intensity of the magnetohydrodynamic (MHD) processes of metal mixing are decreased. To obtain a homogeneous alloy with high accuracy of temperature distribution in the molten bath, power supply of furnace is produced by high and low frequencies currents at different stages of melting [2].

2. Dual-frequency induction melting

In the existing induction melting systems multi-frequency power inductor furnace are implemented [3]. The main mode of heating and melting of the batch mixture is carried out when induction furnace is supplied by the high frequency currents close to the resonant frequency of the load circuit. To supply inductor furnace by low frequency currents it is necessary to reduce the resonance frequency of the load circuit, for example, by switching of series-connected blocks of compensating capacitors in parallel circuit connection to the inductor.

The Otto Junker Co. (Germany) offered the "Multi-Frequenz" technology, which was tested at the MAN Diesel foundry of grey iron in Denmark. Decarburization of metal was made by changing power supply mode of inductor furnace from 250Hz to 125Hz. Studies have shown that the duration of the decarburization process was reduced to 1.5-2 times by increasing the metal speed of movement, thus energy consumption and the cost of production of synthetic cast iron were significantly reduced.

2.1. Changing power supply mode of furnace

Similar multi-frequency induction melting system was developed in the Ural Federal University and the Research and Production "RELTEC" Co. (Russia, Ekaterinburg) whose electric circuit is shown in figure 1 [4]. In this circuit, cascade current inverter (CCI) assembled on thyristors VS1, VS2 and VS3, VS4 is applied, which operates in synchronous mode of switching on the switch of the first and second cascade to the load circuit formed by series-connected capacitor banks (C1, C2) and furnace inductors (I).

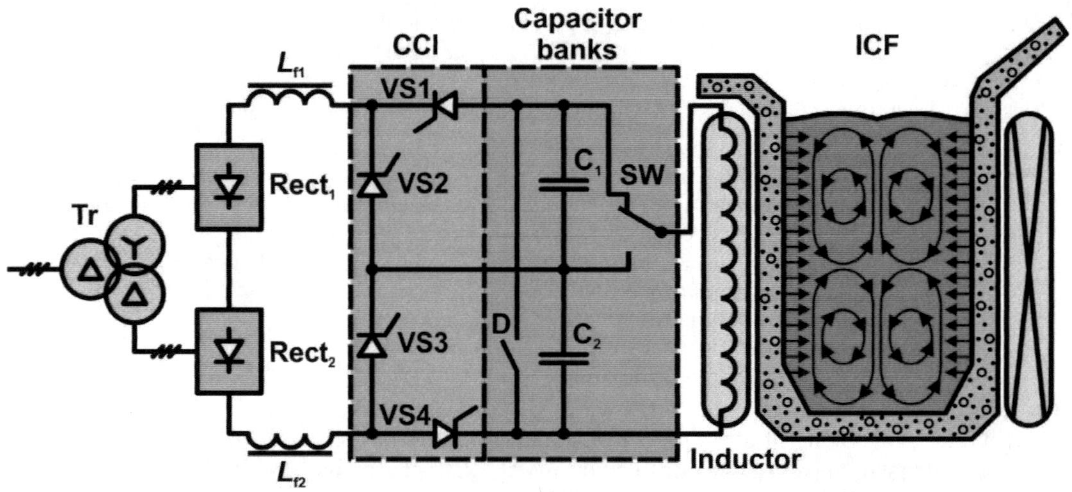

Figure 1. Electrical circuit of the frequency converter with power mode switching of the furnace.

When operating at high frequency, voltage on the inductor is twice higher compared to the bridge circuit of the current inverter, which allows to reduce the level of the inductor current and reduce the active losses in the load circuit proportional to the square of the current. After melting the metal at the stage of technological alloying and decarburizing operations, the switching mode of the inductor's power supply is produced at low frequency by parallel switching of the compensating capacitor blocks by a disconnector (D) and connecting the inductor of the furnace with a switch (SW) to the diagonal of the alternating current of the inverter bridge VS1 ÷ VS4. Thus, the resonant frequency of the load circuit is halved and the voltage on it is also reduced by a factor of two by switching the current inverter circuit from cascade to bridge circuit, which in turn gives an opportunity to interface the load and the frequency converter parameters.

In this power supply system of induction crucible furnace power switch devices (D and SW) must be rated for full current of the inductor, which significantly complicates and increases the cost of the equipment.

Frequency change of inductor current increases the action on the molten metal and the metal speed of movement, but does not change the process of metal circulation in the two toroidal circuits, heat and mass transfer between them is negligible.

2.2. Dual-frequency power supply

To improve the efficiency of metal movement control the furnace inductor is expediently performed as a sectionalized and each section is better to be powered simultaneously with high and low frequency currents, and at different stages of melting it is necessary to be able to change the ratio of high and low frequency currents as well as phase shift.

Power supply of the furnace inductor sections by low frequency currents allows increasing the electromagnetic pressure on the molten bath and the metal speed of movement, and by changing the ratio of the low-frequency current in the inductor sections and their phase shift, control of the direction of metal movement is achieved, which contributes to the expansion of technological capabilities and induction melting efficiency in obtaining exact alloys.

Thus, the simultaneous power supply of the crucible furnace with high and low frequency currents makes it possible to increase the efficiency of induction medium-frequency melting of metals and to expand the technological capabilities of the furnace by controlling the movement of the metal under the action of low-frequency currents. Figure 2 indicates the electric circuit diagram of a dual-frequency converter and induction crucible furnace, in which the inductor consists of two coaxially placed sections having right and left winding of turns in each section. Compensation of the reactive power of the furnace is carried out by high-frequency capacitors (C_{HF}) connected in parallel with the upper section of the inductor, and low-frequency capacitors (C_{LF}) cover both inductively connected sections of the inductor.

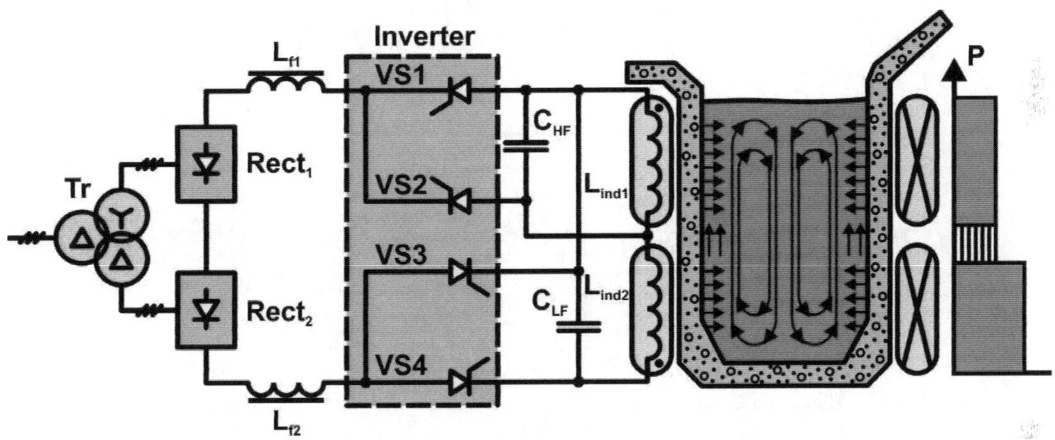

Figure 2. Schematic circuit of a dual-frequency converter and an induction crucible furnace with a diagram of the electromagnetic pressure.

Two-circuit current inverter is connected by means of its output to a low-frequency and high-frequency capacitor. The high-frequency cascade of the inverter (VS1, VS2) operates at higher frequency in relation to normal frequency of the high-frequency load loop formed by the parallel sections of the inductor (L_{ind1}, L_{ind2}) through a low-frequency capacitor (C_{LF}) and a high-frequency capacitor (C_{HF}), therefore it allows commutating thyristors of this cascade.

Low frequency cascade inverter (VS3, VS4) operates synchronously with high frequency cascade at frequency which is in three times lower. In this case the normal frequency of the low frequency load circuit, formed by series connected sections of the inductor (L_{ind1}, L_{ind2}) and the capacitor low frequency (C_{LF}), is to be slightly higher than the frequency of the low frequency of the cascade, and the conditions of switching of its thyristors are provided by a high-frequency component of the voltage on the capacitor low frequency (C_{LF}).

Figure 3 demonstrates diagram of two-circuit current inverter operation. Low frequency cascade forms output current of the low frequency I_{ind} with rectangular shape (see figure 3(a)), which consists of main $I_{1\ ind}$ and third harmonics $I_{3\ ind}$. Low frequency $U_{C\ LF}$ and triple frequency $U_{C\ LH}$ voltages are formed on the capacitor C_{LF}. They are shifted relative to the harmonic components of output current (β_L and β_3). The total voltage U_{CL} on low frequency capacitor (see figure 3(e)) determines the voltage

IOP Conf. Series: Materials Science and Engineering **313** (2018) 012008 doi:10.1088/1757-899X/313/1/012008

at the thyristors of the low-frequency cascade U_{VL}, the diagram of which is shown in figure 3(b). Switching conditions of thyristors are provided by the presence of reverse voltage at the switching time of inverter ϑ_L, the value of which is determined by the mode of high frequency cascade operation.

High frequency cascade forms the triple frequency output current with rectangular shape I_{indH} (figure 3). Main harmonic of that current $I_{1\ indH}$ and voltage $U_{C\ HF}$ (shifted in phase by the angle β_H) at the high-frequency capacitor, determine the power of a high frequency cascade. Low-frequency current branches off high frequency capacitor. Branched current determines the low-frequency component of voltage $U_{C\ HL}$. High and low frequency total current at high frequency capacitor (see figure 3(f)) determines the voltage at the thyristors of the high-frequency cascade U_{VH}. Diagram of U_{VH} is shown in figure 3(d).

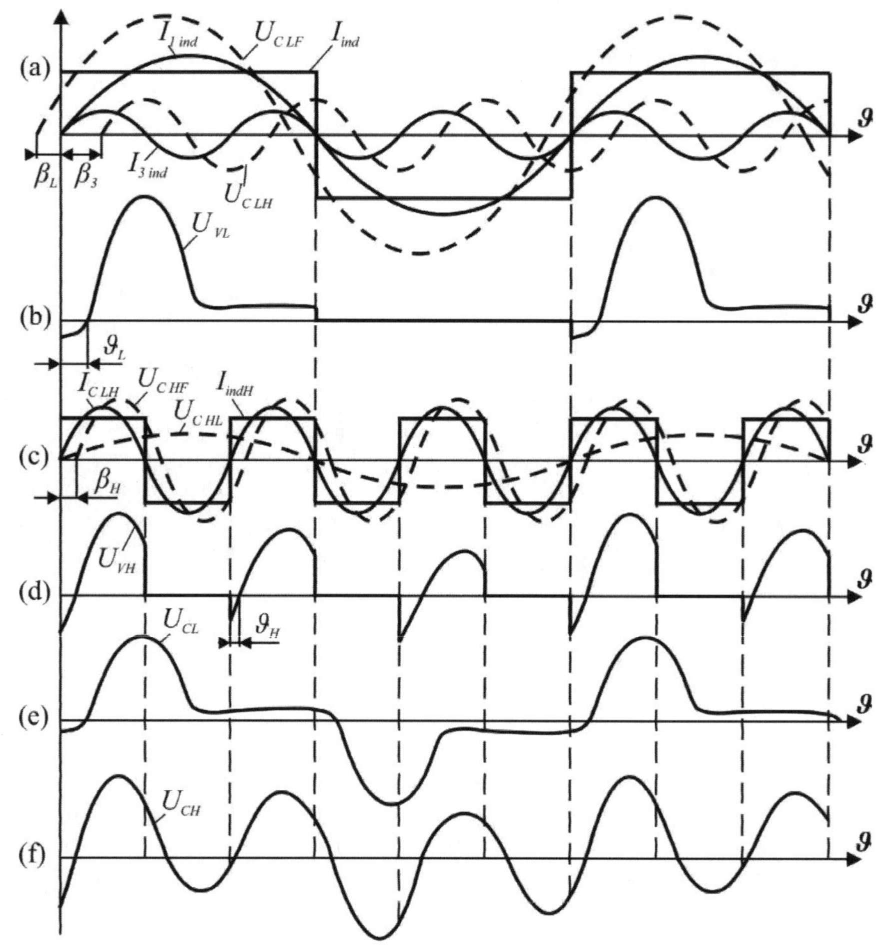

Figure 3. Diagram of two-circuit current inverter operation.

Thyristor switching conditions of high-frequency cascade are due to the presence of reverse voltage at the time of switching and due to recovery angle of thyristors ϑ_H. Value of ϑ_H is determined by the ratio of high and low frequency voltages.

Thus, the high-frequency cascade is leading at the inverter operation, and the low-frequency cascade is the trailing. When the control frequency of the high-frequency cascade increases, the degree of the operating frequency offset from the resonant frequency of the high-frequency load circuit increases, and the operating frequency of the low-frequency cascade, on the contrary, is close to the resonant frequency of the low-frequency load circuit. As a result of control frequency regulating of the leading cascade, the ratio of the low and high frequency current components changes in the inductor sections, and consequently the required levels of the high-frequency and low-frequency current components in the inductor at different stages of the metal melting process can be established. At the stage of heating and melting of the batch mixture, a maximum level of high-frequency current is required, at which the most intense heat release in the metal occurs.

After melting and forming a liquid metal bath, it is advisable to carry out intensive movement and mixing of the metal under the power action of the low-frequency component of the current in the inductor, the value of which should be maximum, during the process of alloying, homogenizing and melt temperature balancing. Figure 2 shows a diagram of the electromagnetic pressure distribution over the height of a metal bath under the influence of low-frequency current components in the inductor sections. The electromagnetic normal pressure exerted by the upper section is slightly lower than that created by the lower section, because the upper section of the inductor is shunted by a high-frequency capacitor (C_{HF}) and the level of low-frequency current in it is lower. In addition, the low-frequency current of the upper section is slightly behind the current in the lower section of the inductor, which is responsible for the effect of tangential electromagnetic forces on the molten metal directed upwards in the inter-winding space.

3. Control of the power

Analysis of inverter electromagnetic processes is advisable to perform by fundamental harmonic method because furnace power load circuits have high quality factor. Figure 4 shows a simplified equivalent circuit of a two-stage current inverter, working on low frequency circuit which in turn is formed by inductor series connected sections of the L_{ind1}, L_{ind2} and low frequency capacitor C_{LF} and high frequency circuit which is represented by parallel connection of capacitors C_{HF}, L_{ind1} inductor section and series-connected capacitor C_{LF} and L_{ind2} inductor sections.

Inverter low-frequency cascade which forms current rectangular shape $I_{ind} = 0,5I_d$ can be replaced by a source of sinusoidal current of the fundamental harmonic

$$i_{(1)LF} = 0,45I_d \cdot \sin \omega t, \tag{1}$$

and also by the source of third harmonic component

$$i_{(3)HF} = 0,15I_d \cdot \sin 3\omega t. \tag{2}$$

High frequency inverter cascade works in cophasal mode with third harmonic component of low frequency current and at frequency which in 3 times higher than frequency of low-frequency cascade. High frequency inverter forms rectangular high frequency current $I_{indH} = 0,5I_d$. The main harmonic of this current can be defined as

$$i_{(1)HF} = 0,45I_d \cdot \sin 3\omega t. \tag{3}$$

Capacitance of compensating capacitors C_{LF} и C_{HF} are selected on the basis of balance between resonant frequency of power circuits ω_{0LF} и ω_{0HF} and operating frequency of cascade inverter. However, to implement frequency control method of high and low frequency current levels $K_I = I_{LF} / I_{HF}$ in the furnace inductor sections it is advisable to install the double ratio of natural frequencies oscillating circuits. If double ratio is installed than deep regulation of low and high currents frequency $K_I = 1 \div 3,5$ is achieved and stabilization of the total power transmitted to load P_Σ is provided at the high and low frequency.

In the proposed circuit of a the two-circuit parallel current inverter without switching devices, it is possible to control smoothly the low-frequency component of the current in the inductor sections of the furnace. At the stage of heating and melting the batch mixture, it is necessary to apply high level of the high-frequency component of the current in the inductor sections, at which the melting process is the most effective. At the stage of technological treatment of liquid metal (alloying, decarburization), it is necessary to increase the intensity of mixing by raising the level of the low-frequency component of the current in the inductor sections. In the proposed circuit, a smooth change of the ratio between low and high frequency component K_I is ensured by the correct choice of the ratio between normal frequencies of the low and high frequency load circuits. Regulation of this ratio is achieved by changing the control frequency of high frequency cascade, f_H and the frequency control of low frequency cascade should be three times smaller than $f_L = 1/3 \cdot f_H$.

Figure 5 demonstrates behaviour of total power transmitted to the load P_Σ, power of low-frequency cascade of P_L, high-frequency cascade P_H and the coefficient K_I. These characteristics are calculated for an 8 tons induction furnace. Characteristics show that stabilization of the total power P_Σ in the regulation of f_H is achieved by a significant change of K_I (about 3 times). This ensures electrical load constancy of converter power equipment elements and rate of melting metal. A significant change of low-frequency current component in the furnace inductor sections at the end of melting process allows to change the character of liquid metal movement and to convert it from a two-circuit toroidal motion to single-circuit toroidal motion throughout the bath volume. In addition, under the influence of low frequency currents metal speed of movement significantly increases, which contributes to acceleration of technological processes of alloying, temperature equalization in the molten metal bath, homogenization of alloy. In the production of synthetic cast iron control of the ratio between low and high frequency components of currents allows to reduce the decarburizing time by several times.

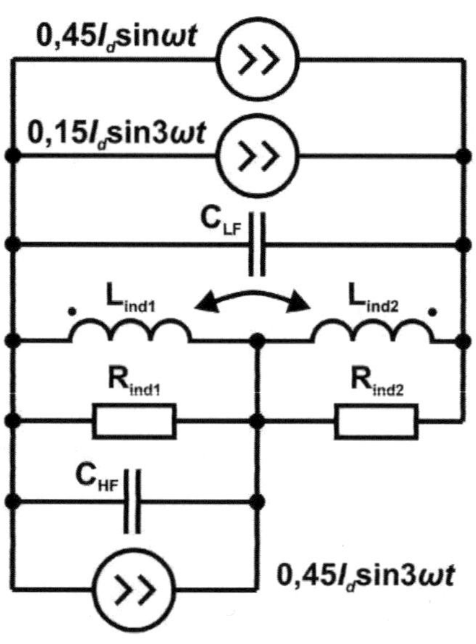

Figure 4. Equivalent circuit of a two-stage current inverter.

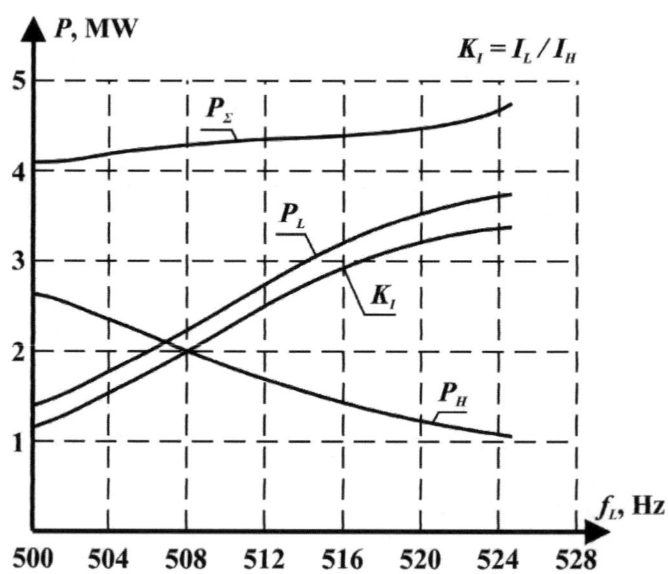

Figure 5. Frequency characteristics of the power supply.

4. Conclusion

Studies of double-frequency melting induction systems of metals have shown that using multi-frequency heating makes it possible to realize flexible control of the heating and metal movement in the furnace bath through organization of necessary heat and force action of electromagnetic field. Thus it is possible to create new technologies for the production of high quality metals and alloys for cost-effective foundries.

Particularly, actual application of innovative technologies of metals induction melting becomes at the present stage of foundry factories development, allowing to diversify maximally foundry production at minimum costs.

References

[1] Gandhewar V R, Bansod S V and Borade A B 2011 Induction Furnace: a Review *Int. J. of Engineering and Technol.* vol 3 **4** 277

[2] Frizen V E, Luzgin V I, Koptyakov A C, Bolotin K E, Tarchutkin N V and Mironov S E 2017 Induction crucible furnace with dual frequency power supply *Proc. Int. Conf. on Electrical Machines, Drives and Power Systems, 15th* (Bulgaria: Technical University of SofiaSofia) pp 423-426

[3] Trauzedal D 2003 Realization of special technological and metallurgical tasks in optimized medium frequency furnaces *J. Russian Foundrymen* **5** pp 20-23

[4] Lusgin V I, Petrov A U and Fatkullin S M 2013 Induction melting and casting complexes for the production of high-quality cast iron *J. Russian Foundrymen* **5** pp 27-30

IOP Conf. Series: Materials Science and Engineering **313** (2018) 012009 doi:10.1088/1757-899X/313/1/012009

Investigation study of geometric dimensions of the magnetic system of the switched-reluctance machine influence on magnetic moment

A Petrushin[1], A Shevkunova[1]

[1] Rostov State Transport University, 2, Rostovskogo Strelkovogo Polka Narodnogo Opolcheniya sq., Rostov-on-Don, 344038, Russia

E-mail: alex331685@yandex.ru; nastya3051990@mail.ru

Abstract. The article deals with the investigation concentrated to optimizing the active part of the switched-reluctance motor with the aim of increasing the value of the average electromagnetic torque. Susceptibility of the average value of the electromagnetic torque to changes of the geometric dimensions of the magnetic system found in the optimization process was set.

1. Introduction

Electromechanical converters of energy (EMC) are the basis of modern industrial production. Constantly increasing requirements to dynamic, the energy and mass-dimensional characteristics of EMC require the improvement of all components and active parts involved in electromechanical energy conversion.

EMC switched-reluctance type having a sufficiently high technical and economic indicators is one of the most promising at present. In the world practice, a switched-reluctance electric machines (SRM), composed of switched-reluctance electric drives (SRD) are used in many fields. The main reason for the expansion of their application is a number of advantages such as design simplicity, reliability, relatively low cost to manufacture and a high energy performance and mass-dimensional indicators. In this regard there is a need to improve their design, in particular magnetic systems with the aim of improving technical and economic indicators. Due to the availability of powerful computers and software, it is possible to improve the active part of SRM through the use of optimization algorithms.

In most cases, three-phase and multiphase SRM are used there some areas where a single-phase SRM application is rational. For example: for household appliances [1], for machines and mechanisms with ventilation load, for operation in generator mode, in the cases where there is requirements no high to the starting point and the major argument is the low cost. The switched-reluctance motor (SRM) in some cases is used as an embedded object in the technological machine, which may entail a number of design restrictions for the magnetic system. This study was conducted in order that the designer had the opportunity to choose certain parameters for optimization that have the greatest influence on the magnitude of the electromagnetic torque.

IOP Conf. Series: Materials Science and Engineering **313** (2018) 012009 doi:10.1088/1757-899X/313/1/012009

2. The objectives of the investigation

The main aim of the study was optimization of single-phase SRM 6/6 magnetic system (fig. 1) according to the criterion of maximum of average value of the electromagnetic torque and the determination of the susceptibility of electromagnetic torque value to the geometric dimensions of the magnetic system changes found in the optimization process.

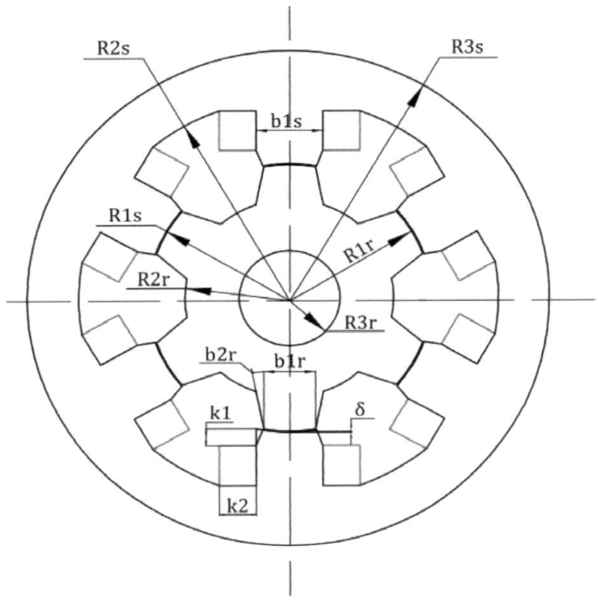

Figure 1. Magnetic system SRM 6/6

3. The solution of the problem

The object of the investigation is SRM (fig. 2), which was manufactured at the plant "Horizont" (Rostov-on-Don) and it was tested as part of the ventilation unit successfully.

Figure 2. Single-phase SRM

Basic dimensions and nominal values are presented in the table 1, 2. As a tool for the automated design and optimization process the program "Optimization of SRM" developed by the authors was used [2]. It had already been used for optimization of SRM with other types of magnetic systems [3–6].

To determine the dynamic characteristics of the single-phase SRM under consideration the known relations [7]

$$\frac{di}{dt} = \frac{1}{L_d(i,\theta)} \cdot \left(u - i \cdot R - \omega \cdot \frac{\partial \psi(i,\theta)}{\partial \theta} \right),$$

$$\frac{d\omega}{dt} = \frac{1}{J} \cdot (M_e - M_s).$$

where i is the phase current, t is the time, L_d is differential phase inductance, u is the phase voltage, R is active resistance of the phase winding, ω is frequency of rotation of the rotor, ψ is flux linkage of phase, θ is the angle of rotation of the rotor, J is reduced moment of inertia of the rotor, M_e is the electromagnetic moment on the shaft, M_s is reduced moment of resistance on the shaft.

The calculation of the electromagnetic torque was performed through a derivative of coenergy [8, 9].

Table 1. Basic dimensions of the studied SRM 6/6

The name of the parameter	Value
The number of teeth of the stator	6
The number of teeth of the rotor	6
Number of phases	1
The packet length of the stator (rotor), mm	50
Outer radius of stator (*R3s*), mm	44,2
The radius of the stator in the grooves (*R2s*), mm	35,5
The radius of the bore of the stator (*R1s*), mm	24
The width of the tooth of the stator (*b1s*), mm	11,2
The angle of the side surface of the tooth of the stator (*b2s*), grad.	0
The air gap (δ), mm	0,2
The outer radius of the rotor (*R1r*), mm	23,8
The radius of the rotor grooves (*R2r*), mm	17,6
The radius of the hole under the shaft (*R3r*), mm	8,5
The tooth width of the tooth of the rotor (*b1r*), mm	8,7
The angle of the side surface of the tooth of the rotor (*b2r*), grad.	11
Coil width (*k2*), mm	6,25
The elevation of the tooth of the stator on the coil (*k1*), mm	3,02
The number of turns, *n*	130

Table 2. The nominal value of the studied SRM 6/6

Name of engine parameters	Value
Power, W	234
Moment, N·m	1,492
Rotation frequency, min^{-1}	1500
The effective value of current, A	4 A
The voltage pulse DC, V	310

IOP Conf. Series: Materials Science and Engineering **313** (2018) 012009 doi:10.1088/1757-899X/313/1/012009

The table 3 with the results obtained in the study is presented below. Optimization was performed in two stages (for a more accurate determination of numerical values of time): first fragments of the magnetic system was optimized by Monte-Carlo method (to avoid finding a local extremum), and then by using of Nelder – Mead method (for the Refine value). Five parameters highlighted in the table 1 were taken as the base as optimizing geometric elements.

Table 3. The results of the optimization

Optimized parameters	The method of Monte-Carlo, mm	The method of Nelder – Mead, mm	The increase of torque (from source), %
	The average torque *M*, N·m	The average torque *M*, N·m	
$b2s$, $b1s$, $R2r$, $b2r$, $b1r$	$b2s$ = 8,35; $b1s$ = 12,59; $R2r$ = 17,86; $b2r$ = 14,84; $b1r$ = 9,92 M = 2,14	$b2s$ = 6,62; $b1s$ = 13,86; $R2r$ = 16,48; $b2r$ = 14,26; $b1r$ = 10,69 M = 2,143	30,37

The Monte-Carlo method is referred to the methods of static testing because of the need for a sufficiently large number of calculations. The definition of the method is to simulate a random variable in order to calculate the characteristics of their distributions. A detailed description of the application of this method as well as its advantages and disadvantages are considered in the work [10].

The method of Nelder – Mead (or deformable polyhedron) is a continuation of the simplex method Spindle, Hexta and Hemsworth. This method applies to deterministic methods of unconditional optimization of functions of several variables and does not use the derivative (gradient) of the function. A detailed and descriptive description of this method is presented in [4].

Are showed in figure 3 for two geometries: the black color shows the geometry of the source engine, and red – optimized. Demonstration of the final result.

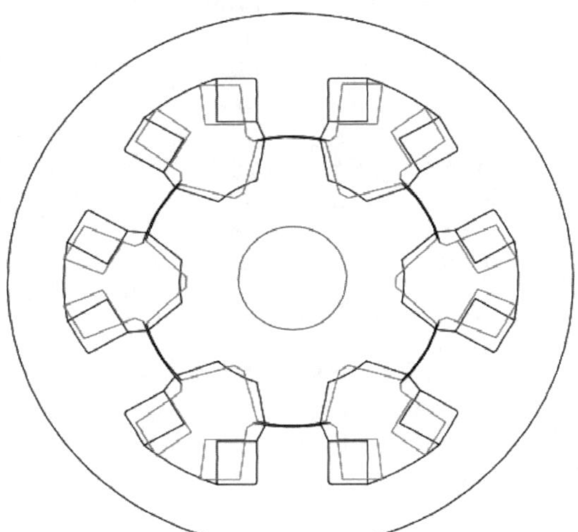

Figure 3. Cross-sections of the original and optimized active part of the SRM 6/6

It can be seen from figure 3 that the zone for the passage of magnetic flux in teeth area of the optimized version of the SRM is wider than the initialone. It is a reason of the decrease of the magnetic resistance, the degree of saturation of the magnetic system and the increase in the average electromagnetic torque. However, further extension of this zone leads to a decrease in the difference between the phase inductance in coordinated and miscoordinated positions that lead to decrease in the average electromagnetic torque.

Due to the fact that the optimization process involves a number of SRM parameters the physical basis of the geometry obtained in the result of the optimization must be coordinated with these parameters. Finite element method, which is the basis of FEMM [11, 12] program allows to estimation so SRM calculations the geometry of the machine obtained after optimization of provides a maximum objective function, in this case contradictory requirements are coodinated.

To determine the sensitivity of the electromagnetic torque to changes in the geometrical quantities of the five parameters (see table 3, column 3) magnetic system SRM 6/6 was carried out the following steps:

1 In the "Optimization SRM" the calculation of the average electromagnetic torque when one of the five parameters (the other four dimensions are without changes) on the value of ± 1–5 % (1 % increments) was carried out.

2 The dependence of the variations of the values of the considered geometrical parameter was built.

According to this investigation we obtained the following values of the average torque and the geometric dimensions (table 4).

Thanks to the data given in table 4, the dependences presented in figure 4. Were obtained reference values of the electromagnetic torque and resizable fragment of the magnetic system are pointed by the red dots on the chairs data.

The dependences obtained are arranged in the order of reducing the effect of size changes on the value of the moment. This is an analysis of data graphs.

Table 4. The average electromagnetic torque when changing the geometric dimensions of the fragments of the magnetic system

	-5%	-4%	-3%	-2%	-1%	**Initial value**	1%	2%	3%	4%	5%
b1r	10.156	10.263	10.37	10.477	10.584	**10.69**	10.79	10.903	11.01	11.117	11.224
M	1.697	1.86	1.95	2.076	2.104	**2.143**	2.114	2.113	2.102	2.086	2.046
b1s	13.177	13.316	13.454	13.59	13.73	**13.87**	14.00	14.147	14.28	14.424	14.563
M	1.785	1.889	2.061	2.088	2.107	**2.143**	2.116	2.085	2.069	2.066	2.038
b2s	6.289	6.356	6.422	6.488	6.554	**6.62**	6.686	6.752	6.818	6.884	6.95
M	1.789	1.94	2.054	2.086	2.105	**2.143**	2.119	2.089	2.075	2.065	2.04
R2r	15.656	15.821	15.986	16.151	16.316	**16.48**	16.64	16.81	16.97	17.139	17.304
M	2.141	2.1415	2.142	2.1425	2.1428	**2.143**	2.142	2.1425	2.142	2.141	2.139
b2r	13.557	13.7	13.842	13.985	14.128	**14.27**	14.41	14.55	14.69	14.84	14.98
M	2.139	2.14	2.1405	2.141	2.142	**2.143**	2.142	2.1423	2.142	2.141	2.14

In figure 4, *a* the dependence of the average electromagnetic torque on the values of the fragment magnetic system b1r in intervals of ± 1–5 % is presented. The graph shows that the optimal point is the point with such coordinates (10,69; 2,143), and any deviation from the value *b*1*r* = 10,69 mm to the left or right leads to the reduction of the moment. The largest slope is observed the at values period –1 % to –5 % b1r. This means that the reduction in the width of the tooth of the rotor (*b*1*r*) may lead to a reduction in the numerical values of moment of approximately 20 %, and increasing the value of *b*1*r* 4,5 %.

Figure 4, *b*, *c* shows the dependence of the width of the tooth and the angle of inclination of the side surface of the tooth of the stator (*b*1*s* and *b*2*s*). The forms of these curves are almost similar to figure 4, *a*. It suggests that the decrease in the values of *b*1*s* and *b*2*s* leads to the greatest reduction of value of time (17 % – *b*1*s* and 16 % – *b*2*s*), rather than increasing the value of this parameter (at the 5 % *b*1*s* and 4,7 % – *b*2*s*). Changes of to dimensions of these parameters have a lower impact on the average time, which is evident from the results obtained at ± 5 % of the size of the considered fragments.

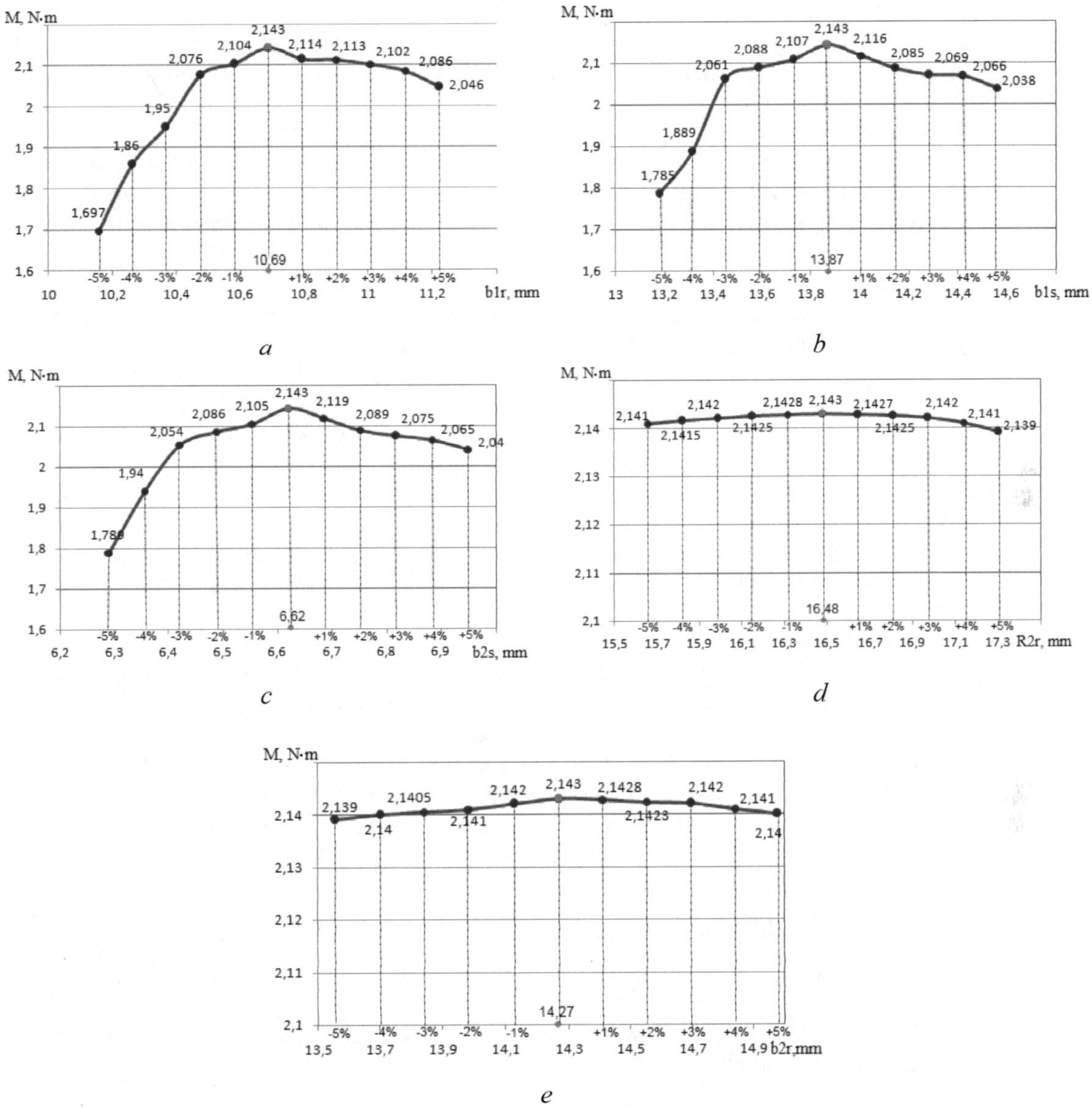

Figure 4. The dependence of the average values of the electromagnetic torque on the variations of the geometric dimensions of the magnetic system SRM 6/6:
a – *b*1*r*; *b* – *b*1*s*; *c* – *b*2*s*; *d* – *R*2*r*; *e* – *b*2*r*

Are presented in figure 4, *d*, *e* the curves of the dependency of the average time from *R*2*r* and *b*2*r*, having a virtually flat form. Are presented it suggests that the variation of data values of parameters of the magnetic system has a minor influence on the formation of the average value of time (± 0,2 %).

4. Conclusions

The result of the optimization algorithm applying was the increase in the average electromagnetic torque by 30,36 % while maintaining the outer diameter and length of the magnetic circuit, which indicates the effectiveness of the application of optimization algorithms in order to improve performance.

The research carried out on the susceptibility of the electromagnetic moment to the change in the geometric parameters of the magnetic system of the object under consideration demonstrates that the greatest influence on the magnitude of the moment is caused by a change in the element $b1r$ (the width of the rotor's tooth).

The value of the torque to the change of parameters $b1s$ and $b2s$ is less sensitive and is the variation of the elements of $b2r$ and $R2r$ very little influenced. These results can be used by the designers in the conditions, when the magnetic system is embedded and the change of any geometrical parameter of the active part is the reason for increase or decrease in operational performance.

In addition, of revealed link between changes of geometric dimensions and the magnitude of the mean electromagnetic moment can be used in the development of the SRM manufacturing technology and the designation of tolerances in design documentation for structural elements.

References

[1] Pakhomin S A, Kolomeitsev F L, Zvezdunov D A, Kolomeytsev V L and Suleymanov W M 1995 Single-phase induction motor type Patent 2040096 Russian Federation (92011863/07)

[2] Petrushin A D, Kashuba A V and Shevkunova A V 2016 "Optimization of the SRM" (2016618039)

[3] Petrushin A D, Shevkunova A V and Kashuba A V 2017 Optimization calculations and experimental studies of the switched-reluctance machine *Internet-journal "Naukovedenie"* vol 9(2)

[4] Petrushin A D, Shevkunova A V and Kashuba A V 2016 Optimization of the active part of the switched-reluctance motor by the Nelder-Mead method *News of the TPU* vol 327(6) pp 83–92

[5] Petrushin A D, Shcherbakov V G and Kashuba A V 2017 Optimization of the magnetic system of a switched-reluctance motor *Izvestiya vuzov. Electromechanics* vol 60(1) pp 20–27

[6] Kashuba V A and Shevkunova A V 2017 Optimization of magnetic system of the switched-reluctance motor when operating in one-pulse mode *Works of RSTU* 2(39) pp 26–31

[7] Wichert T 2008 Design and construction modifications of switched reluctance machines *Ph.D. Thesis* Warsaw university of technology. Institute of Electrical Machines 161 p

[8] Jian L, Ronghai Q, Zhichu C and Yun-Hyun C 2013 Optimization of energy conversion loop in switched reluctance motor for efficiency improvement *J. Electr. Eng. Technol* vol 8(3) pp 565–571

[9] Miller T J E Optimal design of switched reluctance motors 2002 *IEEE Transactions on industrial electronics* vol 49(1) pp 15–27

[10] Shevkunova A V 2016 Design of switched-reluctance motor as a hub in the system variable speed drive with the application of optimization algorithms *Internet-journal "Naukovedenie"* vol 8(4)

[11] Finite Element Method Magnetics 2009 *User's Manual*

[12] Ganji B, Heidarian M and Faiz J 2015 Modeling and analysis of switched reluctance generator using finite element method *Original research article ain shams engineering journal* vol 6(1) pp 85–93

Aspects of intelligent electronic device based switchgear control training model application

Dimitar Bogdanov[1] and Ivaylo Popov[2]

[1] Technical University-Sofia, Faculty of electrical engineering, Electrical power engineering chair, "Relay protection" laboratory, Bulgaria, dbogdanov@tu-sofia.bg

[2] Intertest Ltd., Sofia, Bulgaria, i.popov@intertest.bg

Abstract. The design of the protection and control equipment for electrical power sector application was object of extensive advance in the last several decades. The modern technologies offer a wide range of multifunctional flexible applications, making the protection and control of facilities more sophisticated. In the same time, the advance of technology imposes the necessity of simulators, training models and tutorial laboratory equipment to be used for adequate training of students and field specialists

1. Introduction

In the last decade the control systems of switchgears, substations and power plant facilities were designed on SCADA based architecture and IED (Intelligent Electronic Devices) application.

The "IED" definition is likely over-characterizing the microprocessor multifunctional devices properties, since "artificial intellect" has not yet been implemented. It can be noted, that modern protection and control equipment requires sophisticated preparedness and periodic training from the side of the personnel [1]. The application specifics of IEDs impose requirements for their high reliability [2,3,4,5].

2. Problem

Problems, intended to be solved in some extent with the presented in the paper IED based switchgear control training model: The accumulated experience of tutorial activity in the control and protection area of electrical power facilities indicated some gaps between theoretically presented data (lectures, projects, seminars, etc.) and the practical preparedness of the students to operate industrial equipment. Some main shortages of practical knowledge were found:

- Insufficient preparedness to operate modern digital (microprocessor) based devices and to work with the respective firmware.
- Insufficient knowledge for practical application of control functions in design aspect – interlocks, etc.
- Psychological "barrier" to operate control gear – lack of self-confidence to perform switching sequences.

The tutorial plans of many disciplines needed harmonization with the requirements of the respective companies, hiring the young engineers, upgrade of the experimental base was necessary.

The presented model is aimed to solve part of some educational gaps, and is subject to further "upgrade".

3. IED based switchgear control model design features

For the purposes students training in the Electrical Engineering Faculty of Technical University - Sofia, model based on multifunctional protection and control device (IED) with HMI full size LCD display, able to display the switchgear bay single line diagram (SLD) was designed and constructed. The realization was accomplished in collaboration with company working in the area of design, diagnostics and commissioning of electrical power facilities. The model is intended for demonstrations, training of students and collecting statistical data [1,6,7]. The modeled scheme is applicable for single busbar feeder connection of open – air HV switchyard bay or for compact metal clad MV switchgear as well.

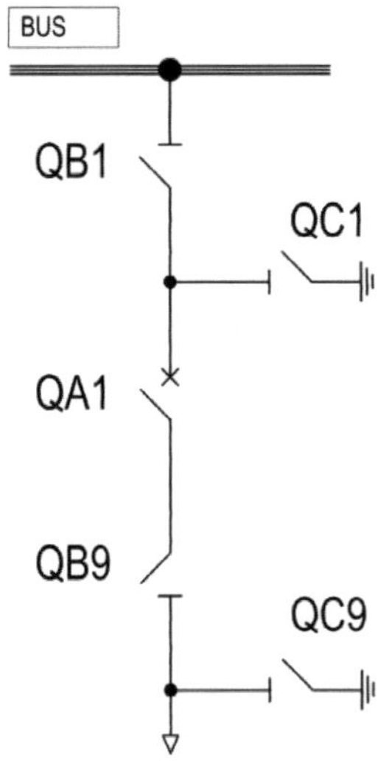

Figure 1. Primary scheme used for model base.

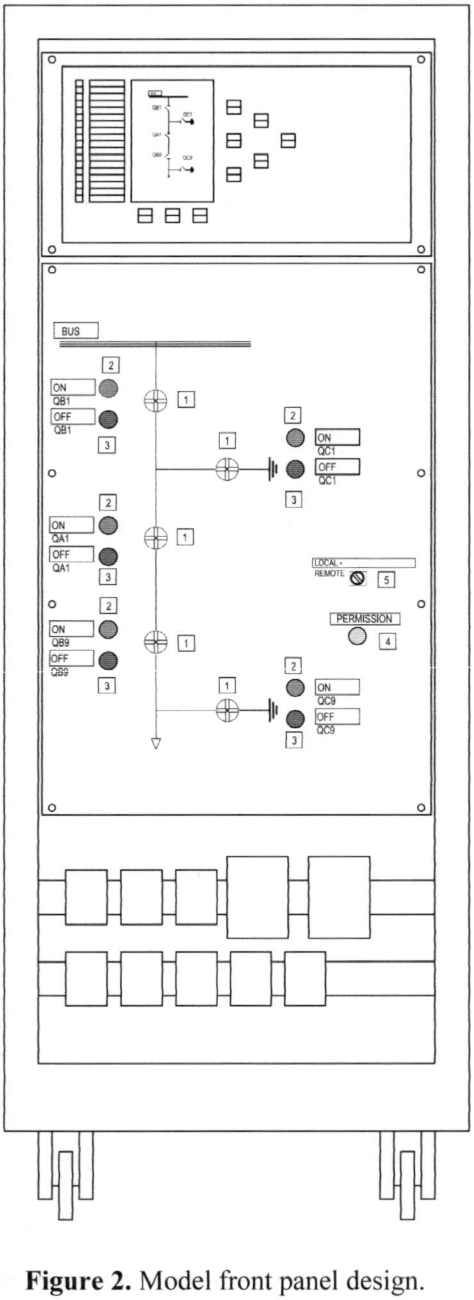

Figure 2. Model front panel design.

IOP Conf. Series: Materials Science and Engineering **313** (2018) 012010 doi:10.1088/1757-899X/313/1/012010

The modeled primary scheme represents a feeder bay, equipped with breaker and all typical commutation apparatus. The design of the model may vary from the "typical industrial design" of some MV schemes, as typically combined commutation devices are used there. Elements of the primary scheme in the scope of the model:

- Feeder circuit breaker;
- Disconnectors;
- Earthling switches.

The modeled scheme is presented in Figure 1. The scheme can be subject of modification – in the IED part functions via the configuration software of the IED. The "hardwire modeled" part – the Bay Local Control Cabinet (BLCC) is realized with physical elements and for functional modification changes shall be made in the scheme wiring.

4. Model design rules

In the practical "field" situation a commutation device, part of primary scheme – breaker, disconnector can be controlled from 4 locations:

- Substation control room via operator's SCADA control station, typical "control tool";
- Protection and control panels via HMI of bay control unit (BCU), used typically during commissioning maintenance, etc.;
- Bay local control cabinet (BLCC) – the existence of BLCC depends on the particular company requirements. If foreseen – used typically for local control during commissioning, in emergency cases, maintenance, etc.;
- Local control buttons on the commutation apparatus control mechanism casing, typically used for testing during commissioning, maintenance.

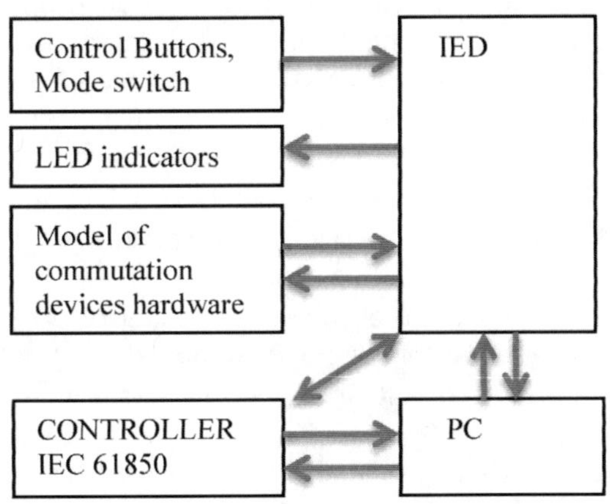

Figure 3. Model functional scheme.

Another possible level of control is the remote dispatch control, applicable for larger and non-manned facilities – typical from dispatch center over communication protocols like IEC 60870-5-101 and IEC 60870-5-104. For the purposes of the training model at the present state the BLCC and HMI of IED devices levels have been realized.

For the particular case, the model hardware structure has 3 main sections:

- IED based control part (multifunctional device resources: protection, control, signalization, data acquisition, communication to PC/controller);

IOP Conf. Series: Materials Science and Engineering **313** (2018) 012010 doi:10.1088/1757-899X/313/1/012010

- Control panel, simulating the BLCC (command buttons, mode switch, LED position indicators);
- Breaker model and accessories (bi-stable relays, auxiliary relays, time relays and elements to simulate the breaker and other commutation devices drive mechanisms and their auxiliary block-contacts).

The model front panel design is presented in Figure 2. The arrangement of all components was designed in one 19" rack cabinet with access to all the components from the front and rear side. The logic has been hard-wired between the IED, control panel and auxiliary elements.

At next stage it is foreseen the BLCC simulating control panel to be reconfigured as detachable unit, remotely located with integrated controller and fiber – optic connection to IED.

The currently constructed functional specifics and the future extensions of the technical solution in limited extent may deviate from the "industrial applications", but the designed structure was targeted mostly to provide access points for students and options to simulate secondary circuit failures, to test Breaker Failure Protection (BFP), autorecloser function, etc.

The BLCC was fitted with additional push button for "permission". The permission button must be kept pressed simultaneously with the respective command button (Close / Open) in order to make the intended commutation operation possible. The functionality was foreseen to avoid erroneous commands by inadvertent button pressing.

The functional scheme (the data/command exchange scheme) of the model is presented in Figure 3.

5. Method of application of the model

The typical tasks which the students must accomplish, performing laboratory practices:

- Check the operation functionality of the model, test interlocks between the commutation devices;
- Test of the commutation devices permitted switching sequence: circuit breaker, disconnectors and earthing switches. The typical performed sequence is taking the breaker bay into "safe" position with all pertaining disconnectors opened and respective grounding switches closed (or reconnection of the feeder to the bus);
- Test of the commutation sequence of grounding switches – check of the possibility for incorrect grounding of live parts;
- Transfer of control mode via the Local / Remote switch on the BLCC panel;
- Verification the configuration of the BCU IED – matrix of signals, etc.;
- Test of protection functions;
- Test of BFP function – option to change the "breaker response time" with modification of the time relay setting, used for breaker operation modeling;
- Overall IEC61850 messages (GOOSE messages) readings with the respective signals verifications;
- Configuration and testing the communication to dispatch center (virtual);
- Learning the structure and the specifics of different communication protocols.

The model allows "errors"/"faults" to be simulated with the IED firmware settings changes and/or wiring changes in the hardware part. The collected information for "personnel response" can be statistically analyzed in different aspects. The response of the "operator" can be estimated in respect to:

- Time to perform successfully a commutation sequence;
- Time to respond to fault conditions (simulated with IED testing device to trigger protection function);
- Time to perform successfully a modification in software / detect and clear a mistake in configuration;
- Time to download records from IED.

In order to estimate ambient conditions impact on the personnel response, the following variants are possible:

- Ambient light variation / low level of lighting;
- Presence of noise source in the surrounding area (alarm siren);
- Presence of other personnel in the vicinity of IED HMI, BLCC.

Definitely, other possible situation can occur "on the field", in the paper is presented limited scope of disturbing factors, related to the switchgear ambience. The performed experiments were organized as a set of switching sequences, accomplished by group of students.

Table 1 presents the results for experimental operation of the model, for several cases of ambient conditions. One series of operations was performed in normal conditions (without "disturbances" for the operator) and the other three series were made with presence of disturbance factors.

Table 1. Results of model experimental operation

	Disturbance factor			
n – number of persons	-	Light	Sound	Attendance
in per cent of total.	n	n	n	n
In the expected Norm t	42.86	28.57	39.29	32.14
In Norm t +05%	17.86	25.00	17.86	21.43
In Norm t +10%	11.00	14.29	14.29	14.29
In Norm t +15%	7.14	10.71	14.29	7.14
In Norm t +20%	10.71	17.86	7.14	10.71
In Norm t +25%	7.14	0.00	3.57	7.14
In Norm t +30%	3.57	3.57	3.57	7.14

The expected time for correct accomplishment of the task was assumed as sum of the operator's switching actions plus the "own time" of the respective devices (commutation apparatus).

$$t_{SO} = t_{OA} + t_{DM}, \quad (1)$$

where: t_{SO} – time for switching operations, t_{OA} - time for operator's actions, t_{DM} - time for operation of the commutation devices drive mechanism (operator independent, design characteristic).

The time for operator's actions can be regarded as sum of the time for accomplishment of the task in normal conditions plus additional time delay caused by the respective disturbance:

$$t_{OA} = t_{OA\,norm} + t_{OA\,ad} \quad (2)$$

where: t_{OAnorm} – time for switching operations in normal conditions (without disturbance), $t_{OA\,ad}$ – additional time imposed by disturbing factor.

The sensitivity of the "successful mission time" in respect of the interfering factors can be calculated as:

$$S_{TOA} = \frac{t_{OA} - t_{OAnorm}}{t_{OAnorm}} \quad (3)$$

In case of combination of the disturbances, the importance of the different impacts on operator's response, can be weighted and combined in common expressions for estimation of the response time.

$$t_{OA} = \sum_{i}^{n} C_{DFi} \cdot t_{OAnorm} \quad (4)$$

where: C_{DFi} – coefficient of importance for the impact of disturbing factor "i" (for the particular case n = 3).

Preliminary experiments of combined ambient area disturbances indicated, that such "combined impact" situation may lead not only to delay in correct response, but to incapability of finalizing successfully a task. The results presented in Table 1 are for experiments accomplished with a single disturbing factor.

IOP Conf. Series: Materials Science and Engineering **313** (2018) 012010 doi:10.1088/1757-899X/313/1/012010

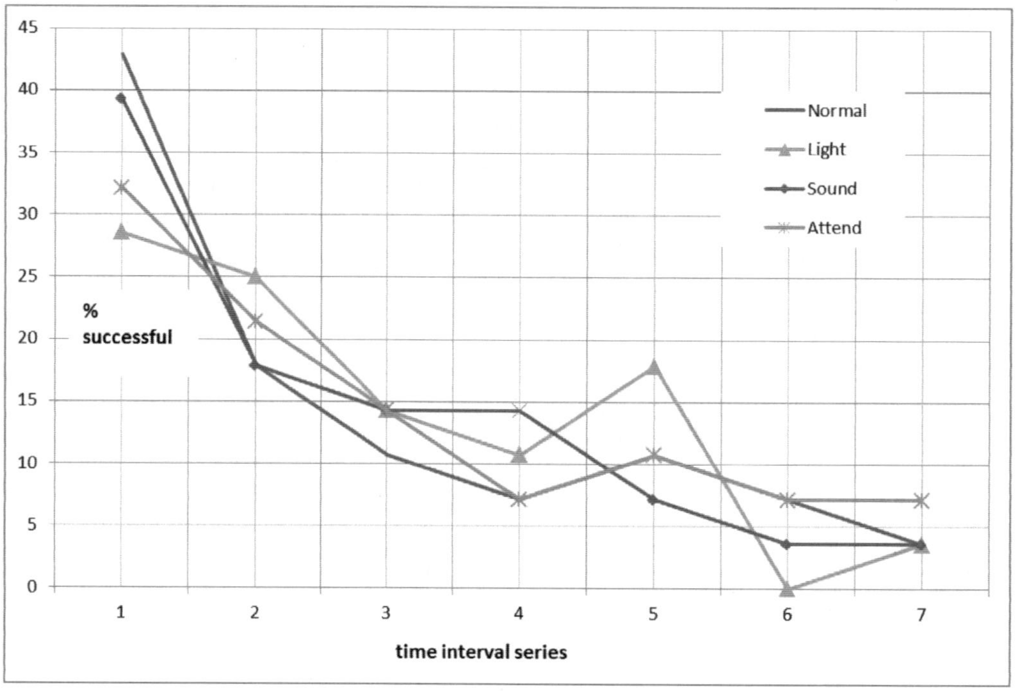

Figure 4. Graphic representation of model experimental operation results.

The conditional probabilities of task successful accomplishment can be defined, if the probabilities for occurrence of interfering factors are known. Such conditional probabilities can be used for prognosis of "failure" or "success" scenarios for particular conditions.

For correct estimation of the results related to operation of real control panel – for instance BCU in HV switchgear, the probabilities of interfering conditions shall be precisely identified, whether they can occur simultaneously and if there are in mutual dependences. Incorrect assumption of independent or mutual excluding factors may distort the assessment probabilistic results.

6. Conclusions

The performed experiments with the model, operated by students proved that it can be used for training purposes and in the same time to collect statistical data for the response of operators, impact of ambient conditions on operator's actions and demonstration of IEC61850 based communication. The procedures for statistical data processing are under development and will target obtaining qualitative and quantitative indices for dependence of the "human factor" in equipment control in respect of the site conditions [8,9,10,11].

The performed tests indicated, that for the particular case the ambient light caused more significant effect on the "operator's response", compared with other disturbance factors. For the particular model – it can be explained with the design specifics: the position indicators are illuminated, but the labeling is "passive" – no back light.

The model in combination with experiments related to ambient work conditions can be used for improvement of control panel's ergonomic design and in the same time to evaluate the impact of the operator working conditions on the probability for successful task accomplishment [11,12,13].

The facility operability dependences in respect of ambient conditions (levels/values of parameters characterizing the equipment ambience) can help finding optimal design towards working environment. The applicable regulations for ergonomic design and safety shall also be taken into account for particular design final decisions.

IOP Conf. Series: Materials Science and Engineering **313** (2018) 012010 doi:10.1088/1757-899X/313/1/012010

Further extension of the model is foreseen for creating a second BCU model for different bus configuration and link to larger existing switchgear model [12]. Combination with primary injection of current in physical busbar system model is also estimated [14]. The option for the students to analyze records of simulated "primary" faults can improve further the training process for response to emergency situations and general level of education in the area of electrical power engineering [14,15].

References

[1] Ching-Lai Hor1, Peter A. Crossley, Knowledge Extraction from Intelligent Electronic Devices; *Transactions on Rough Sets III 82-111 p., January 2005.*

[2] K. Stouffer, J. Falco, K. Scarfone, Guide to Industrial Control Systems (ICS) Security; Recommendations of the National Institute of Standards and Technology, June 2011.

[3] S. Brahma, J. De La Ree, Vice-Chairman, J. Gers, A. A. Girgis, S. Horowitz, R. Hunt, M. Kezunovic, V. Madani, P. McLaren, A. G. Phadke, M. S. Sachdev, T. S. Sidhu, J. S. Thorp, S. S. Venkata,Chairman, T. Wiedman, "The Education and Training of Future Protection Engineers: Challenges, Opportunities, and Solutions," *IEEE Transactions on Power Delivery*, pp. 538-544, Vol. 24(2), 2009.

[4] European Network and Information Security Agency, Protecting Industrial Control Systems, Annex I: Desktop Research Results [Del. 2011-12-09], ENISA, 2011.

[5] European Network and Information Security Agency, Protecting Industrial Control Systems Recommendations for Europe and Member States [Del. – 2011-12-09] ENISA, 2011.

[6] Rangelov Y., N. Nikolaev, New approach for laboratory exercises for students in major electric power engineering; *Proceedings of SIELA2016 conference: Bourgas, Bulgaria, 29 May-01 June 2016: 268 - 274 p.*

[7] Rangelov Y., N. Nikolaev, M. Ivanova, The IEC 61850 standard — Communication networks and automation systems from an electrical engineering point of view; *Proceedings of SIELA2016 conference: Bourgas, Bulgaria, 29 May-01 June 2016: 274 - 278 p.*

[8] Paralleling Switchgear: Documenting the Sequence of Operation with Charts, KOHLER POWER SYSTEMS, www.KohlerPower.com 2016.

[9] Blackout Tracker United States: Annual Report 2016, EATON, https://switchon.eaton.com/blackout-tracker, 2017.

[10] Sahoo P., Probability and mathematical statistics; University of Louisville, 2013.

[11] Shappell S. A., D. A. Wiegmann, The Human Factors Analysis and Classification System–HFACS; Office of Aviation Medicine Federal Aviation Administration, 2000.

[12] Todorov D., A. Krumov, A. Ovcharov, K. Boev, Simulator for operative command panels; *Energetika magazine, vol. 4, 2016, Sofia, 44-50 p. (In Bulgarian).*

[13] L.D. Feisel, A.J. Rosa. The Role of the Laboratory in Undergraduate Engineering Education. *Journal of Engineering Education, pp. 121-130, Jan. 2005.*

[14] Tatyana R. Radeva, Iv. S. Yatchev, D. N. Karastoyanov, N. I. Stoimenov, S. D. Gyoshev. Coupled Electromagnetic and Thermal Field Modeling of a Laboratory Busbar System. *World Academy of Science, Engineering and Technology International Journal of Electrical and Computer Engineering Vol:8, No:9, 2014 1393-1397 p.*

[15] J. Ren, M. Kezunovic. Modeling and Simulation Tools for Teaching Protective Relaying Design and Application for the Smart Grid. *Proceedings of the International Symposium "Modern Electric Power Systems (MEPS)", 2010.*

IOP Conf. Series: Materials Science and Engineering **313** (2018) 012011 doi:10.1088/1757-899X/313/1/012011

Modelling of degradation processes for refractory metallic heating elements of vacuum resistance furnace

M Ya Pogrebisskiy*, A I Bodarev, E F Salmanova and A S Bulgakov

Department of Electric Supply of Industrial Enterprises and Electrical Technology, National Research University "MPEI". Krasnokazarmennaya Street 14, Moscow, 111250 Russia

*PogrebisskiyMY@mpei.ru

Abstract. The mechanism of degradation of the vacuum resistance furnaces heating elements manufactured of pure refractory metals (primarily, tungsten or molybdenum) is analysed. The algorithm of expected life calculation for such heating elements at various values of temperature and vacuum (residual pressure) in the furnace is developed and is programmatically realized. Dependences of expected life on diameter (thickness) of the heater, temperature and pressure are received. The temperature control system for the high-temperature vacuum furnace in the indirect parameter of heaters resistance taking into account their degradation is offered.

1. Introduction

Vacuum electric resistance furnaces are used in mechanical engineering, chemical technology and other industries for carrying out a number of technological processes (heat treatment, agglomeration from powders, including refractory metals powders, roasting of ceramics, receiving carbides, restoration, refinement, a research of properties of materials, etc.). Heating elements medium-temperature (working temperature of 1200–1700 °C) and high-temperature (1700–3000 °C) vacuum furnaces are often made of refractory metals, in medium-temperature furnaces molybdenum or alloys of molybdenum with tungsten (MV-30 and others), in high-temperature – tungsten is generally used.

Forecasting of heating elements life time is important both at design of resistance furnaces, and when planning their operation. In vacuum furnaces at high temperatures and a high vacuum heaters expected life is defined, first of all, by evaporation of their material, at rather low temperatures and an average and low vacuum oxidation by residual oxygen with the subsequent evaporation of oxides prevails. Empirical dependences of material loss rate (due to evaporation and oxidation with the subsequent evaporation) on heater temperature and residual pressure in the furnace camera are given in references [1] for refractory metals, however the reasonable technique allowing to count expected life for heaters of a different form and the sizes in various temperature conditions was absent so far. Especially for refractory metallic heaters in a vacuum there were no similar convenient for use in engineering practice of dependence of expected life on the size of section of the heater and temperature, to those which are well-known for heaters from the high-resistance alloys (nichrome, etc.) operating in the air environment [2, 3].

2. The heater degradation processes analytical model

In references [1] dependences of heaters (tungsten and molybdenum) material loss rate, Q (kg/(m^2·s)), on heater temperature (in range up to 2800 °C for tungsten and to 2000 °C for molybdenum) and pressure in the furnace camera are given (in the range from 10^{-2} to 10^{-5} Pa). However these dependences aren't enough for forecasting of real heater life time.

First, in process of heater material ablation its surface decreases, specific superficial power at the invariable power consumption increases that leads to the heater temperature increase. Thus, heater temperature doesn't remain invariable at the reproduced technological process, that is at an invariable furnace temperature (the temperature of internal heat-shield near which the sensor of temperature regulation system most often is installed) and at constancy of power consumption. With increase of heater temperature during the furnace operation the material loss rate with other things being equal increases that leads to acceleration of the heater degradation.

Secondly, the material loss rate, Q (kg/(m^2·s)), is the quantity carried to the surface area therefore because of heater surface area reduction due to degradation the rate of reduction of its section is changeable even at a constant material loss rate. Most this effect is shown at heaters round (a bar, a wire) sections.

The developed analytical model of the heater degradation process considers such factors as dependence of material loss rate on temperature and residual pressure in furnace camera, increase of specific superficial power and, therefore, heater temperature at reduction of his section, need of increase in supply voltage at reduction of section for maintenance of power and, therefore, the set furnace temperature, dependence of specific electric resistance of material of the heater on temperature.

The mass of material of the heater of round section lost in time interval $\Delta\tau$, is

$$\Delta m = QF_S\Delta\tau \quad (1)$$

where Q – the material loss rate (kg/(m^2·s)) determined on to the above-stated references in dependences on temperature and pressure;

F_S – heater surface area, equal πdL, where d – diameter of the heater, L – its torn length.

The volume of the lost material

$$\Delta V = \Delta m/\gamma \quad (2)$$

where γ – density of heaters material.

New volume of the heater

$$V^* = \frac{\pi d^2}{4}L - \Delta V \quad (3)$$

New diameter of the heater

$$d^* = \sqrt{\frac{4V^*}{\pi L}} \quad (4)$$

Heater temperature in operation is

$$t_H = 100 \cdot \sqrt[4]{\frac{P}{CF_S} + \left(\frac{T_F}{100}\right)^4} - 273, \text{°C} \quad (5)$$

where P – the power consumed by the heater;

T_F – absolute (in Kelvin) furnace (internal heat-shield) temperature;

C – the provided radiation heat exchange coefficient, defined taking into account coefficients, used when the heaters computation and reflecting construction of the heater, its pitch, the given blackness level and ratio of the sizes of the furnace camera and heated products (charge) [2].

Thus, diameter of the heater d^* and its temperature t_H for a present timepoint are calculated. Then the material loss rate and mass of lost material for the following interval of time $\Delta\tau$ are defined (in accordance with (1)), and diameter of the heater and its temperature (in accordance with (2)-(5)) for the following timepoint are calculated, and so on.

With reduction of section of the heater due to its degradation the resistance of the heater increases, and to support the power which is allocated in the heater and, respectively, furnace temperature with constants, the regulator of temperature increases voltage given on the heater. Voltage can be calculated as

$$U = \sqrt{P\rho \frac{4L}{\pi(d^*)^2}} \tag{6}$$

where ρ – specific electric resistance of heater material, defined taking into account known temperature dependence.

The expressions similar (2)-(6), are received for heaters of rectangular section (a tape, a plate).

Further at improvement of model it is planned to provide taking note of gas emission of elements of the furnace design and the charge and vapours of oil in working space on degradation of the heater and also to adapt model for the heaters operating not in a vacuum, and in various gases (argon, etc.).

The realized algorithm developed on the basis of the offered mathematical model and programmatically allows to calculate the current values of the heater diameter (or thickness), its temperature and required supply voltage for any timepoint of operation.

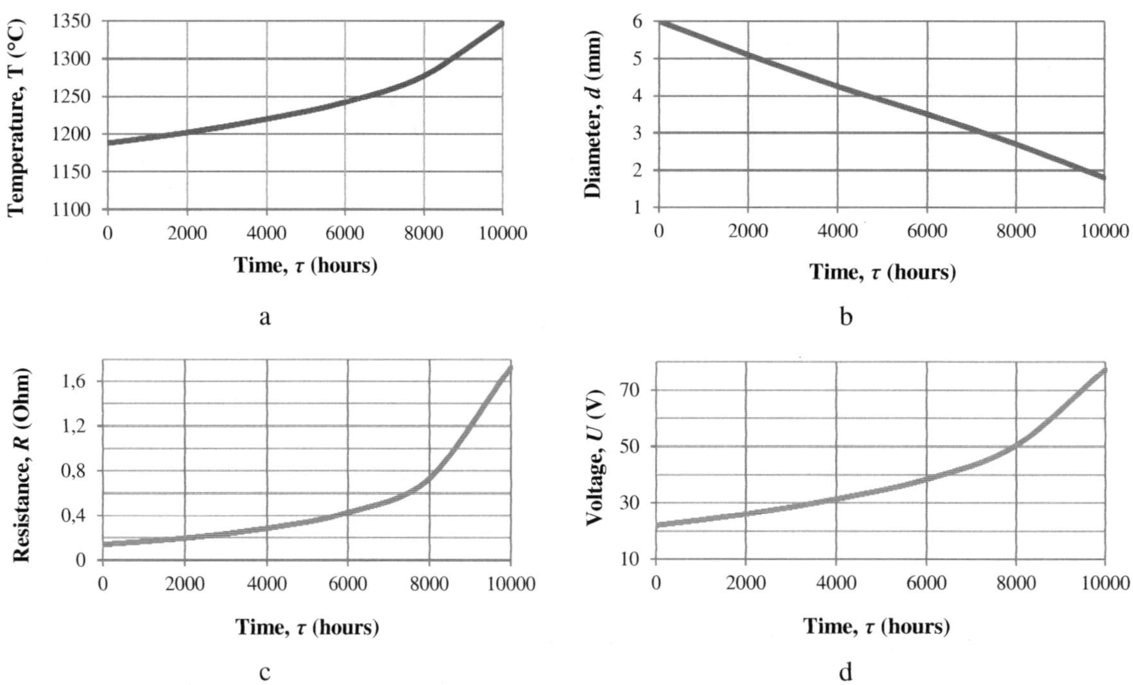

Figure 1. Change of temperature of the heater (a), its diameter (b), resistance (c) and required supply voltage (d) during operation of the molybdenic heater of round section with a initial diameter of 6 mm at a furnace temperature 1100 °C.

IOP Conf. Series: Materials Science and Engineering **313** (2018) 012011 doi:10.1088/1757-899X/313/1/012011

3. Research of heater degradation processes on mathematical model

With use of the developed mathematical model calculations of expected life for heaters of various configurations are made under various conditions (temperature, pressure).

Figures of 1a-d show an example of change of temperature, diameter, resistance of the heater of round section and required supply voltage depending on operation time. The molybdenic heater with an initial diameter of 6 mm used in the vacuum furnace of chamber type with the size of working space 0.3× 0.4 × 0.3 m was considered, furnace temperature that is actually internal heat-shield, 1100 °C.

In the analysis of the given dependences it is necessary to consider that in practice the termination of heater life time should be considered the moment when heater resistance has increased (in connection with reduction of heater section) by 20-30% in comparison with a reference value as with a further growth of resistance a stock on voltage of the power supply will be insufficiently for compensation of this growth°.

Unlike heaters of round section, at flat heaters increase in heater temperature in process of degradation is with other things being equal expressed slightly.

The calculated dependences of expected life of molybdenic or tungsten heaters of round and rectangular section on the initial diameter (thickness) of the heater, furnace temperature and residual pressure in the camera are practically useful. So, at a temperature of 1600 °C and vacuum of 10^{-2} Pa calculated expected life for the heater makes of a molybdenic wire with a diameter of 6 mm is 300 hours that will be coordinated with operating experience of furnaces. Examples of such dependences are given in Figure 2.

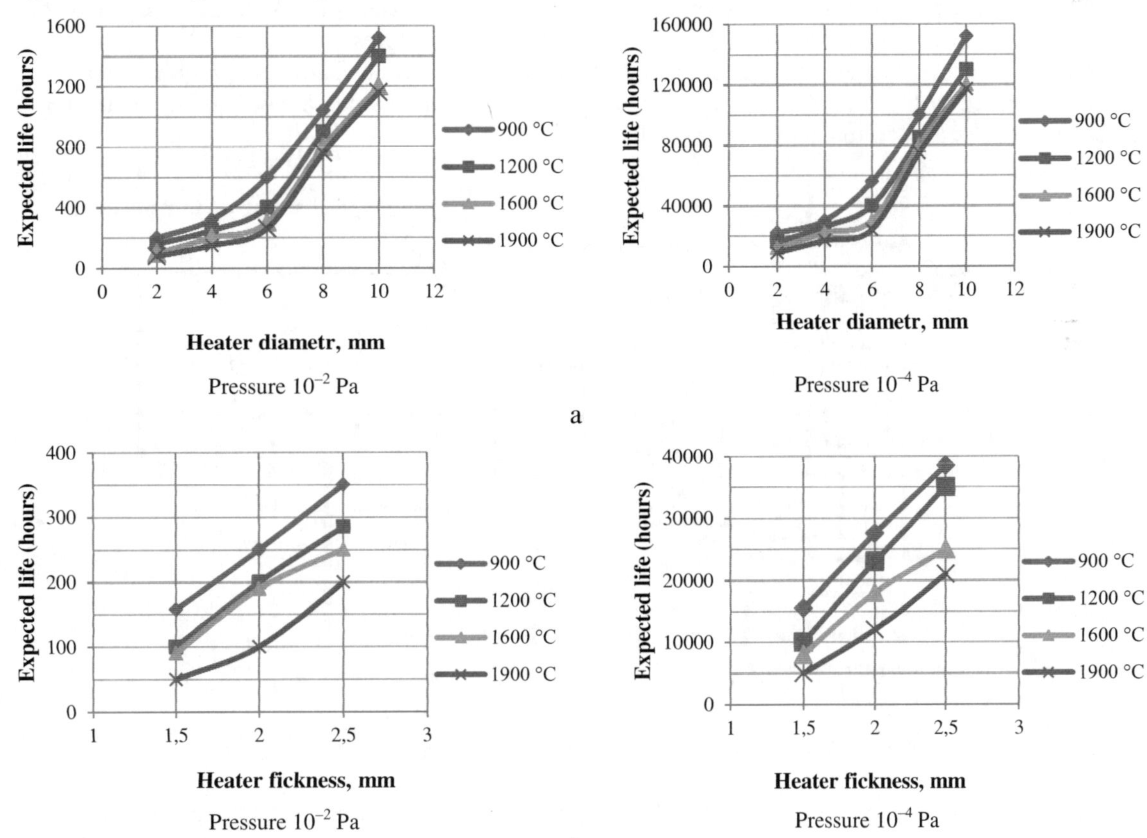

Figure 2. Dependences of expected life (before increase in resistance by 30%) for molybdenic heaters of round (a) and rectangular (b) section on the initial geometrical size (diameter or thickness) at various temperatures and pressure in the furnace camera.

Comparison of expected life for heaters of round and rectangular section shows that with other things being equal (with identical the volume of working space, temperature, pressure and power) heaters of round section serve in furnaces significantly (by 2.5–3 times) longer.

4. Automatic furnace temperature control in the indirect parameter of heaters resistance adjusted for degradation

At temperatures over 1800 °C characteristic of high-temperature vacuum resistance furnaces, the firmness of traditional temperature sensors (thermocouples) is very small. Use as temperature sensor of radiation pyrometers in vacuum furnaces with heat-shields is almost impracticable due to the lack of direct visibility of working space. Thus, in high-temperature vacuum furnaces temperature control should be carried out by the feedforward principle that negatively affects regulation accuracy.

Use of the principle of control in the indirect parameter of heaters resistance will allow to improve the accuracy of temperature regulation in high-temperature furnaces. Temperature calculates on the known functional dependence of specific electric resistance of heater material (tungsten) ρ on temperature t. In turn, specific resistance is defined (on the example of the heater of round section) as

$$\rho = \frac{U_a}{I_a} \cdot \frac{\pi d^2}{4L} \tag{7}$$

where U_a, I_a– the measured values of active components of first harmonicas voltage and current.

Thus, calculation of specific resistance requires knowledge of section (diameter or thickness) of the heater, and reduction of section owing to degradation during heater operation leads to emergence of an error in determination of specific resistance and, therefore, temperatures.

Improvement of accuracy of temperature determination in indirect parameter requires corrective action on reduction of section of the heater during operation. Such amendment can be calculated on the basis of the presented mathematical model.

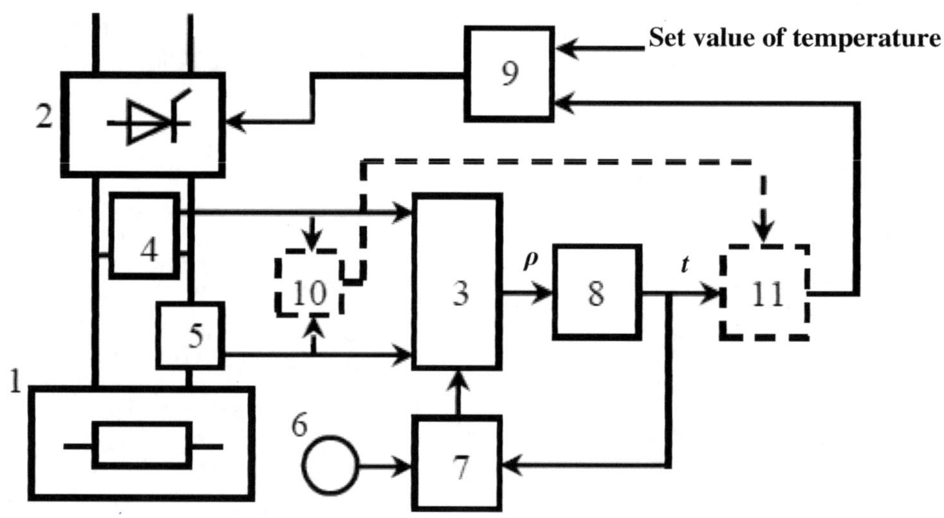

Figure 3. The function chart of temperature control system in the indirect parameter of heaters resistance

1 – vacuum resistance furnace; 2 – thyristor voltage regulator; 3 – block of determination of specific resistance; 4 – voltage sensor; 5 – current sensor; 6 – pressure sensor; 7 – block of corrective action; 8 – block of heater temperature calculation; 9 – temperature regulator (controller); 10 – block of power determination; 11 - block of furnace (or charge) temperature calculation

The function chart of temperature control system in the indirect parameter of heaters resistance is submitted in Figure 3. Object of control is the high-temperature vacuum resistance furnace 1, as the actuation unit of temperature control system traditionally serves the thyristor voltage regulator 2. Block of determination of specific resistance 3 counts specific resistance of heaters material on the basis of signals of voltage sensor 4 and current sensor 5 taking into account the amendment on heaters degradation counted by the block of corrective action 7. An algorithm of calculation of the amendment by the block 7 is under construction on the basis of the offered model of heaters degradation processes. The block 7 defines instant value of heaters material losses rate according to temperature and pressure (pressure sensor is position 6) also counts (in real time) the current diameter (section) of the heater which is considered by the block of determination of specific resistance 3. The block of heater temperature calculation 8 defines the current value of temperature according to the known dependence $t(\rho)$. This value of temperature is used by the traditional temperature regulator (for example, PID-controller) 9 as a feedback signal and also transferred to the block of corrective action 7 for definition instant material losses rate.

In addition using the block of power determination 10 and the block of furnace (or charge) temperature calculation 11, it is possible to count heater temperature in furnace temperature or charge temperature on the basis of a radiation heat transfer lows, knowing power consumption.

Blocks of determination of specific resistance, corrective action, power determination and temperatures calculation can be realized on the basis of the microprocessor control unit.

We will note that the current sensor, as a rule, is already used in the existing control systems of vacuum furnaces with refractory metallic heaters due to the need of current restriction at warming up of the heater from a cold state.

References

[1] Marmer E N 2007 *Materials for high-temperature vacuum installation* (Moscow: Fizmatlit) 152 p
[2] Altgauzen A P, Nekrasova N M and Gutman M B 1980 *Electrothermal equipment: Reference book* (Moscow: Energy) 416 p
[3] Zukov L L, Plemyannikova I M, Mironova M N, Barkaya D S and Shumakov Yu V 1985 *Alloys for heaters* (Moscow: Metallurgy) 146 p
[4] Pitchard J 2007 *Hot-zone design for Vacuum furnaces* (Ontario)
[5] Pogrebisskiy M Ya and Batov N G 2011 *Materials for electric resistance furnaces* (Moscow: MPEI) 92 p

Improving the engineering-and-economical performance of ore-thermal electric furnaces in the smelting of silicomanganese

V P Kondrashov[1,*], **M Ya Pogrebisskiy**[2,**], **A G Lykov**[1], **V L Rabinovich**[3,***] **and A S Bulgakov**[2]

[1]Eltertechniks, Ltd. Novoostapovskaya Street 66, Moscow, 115088 Russia
[2] Department of Electric Supply of Industrial Enterprises and Electrical Technology, National Research University "MPEI". Krasnokazarmennaya Street 14, Moscow, 111250 Russia
[3] Research and production centre "Energotechnologiya". Nizegorodskaya Street 70-2, Moscow, 109052 Russia
[*] etertechniks@yandex.ru; [**]PogrebisskiyMY@mpei.ru; [***]npfenergo@yandex.ru

Abstract. Ways of increase of ore-heating electric furnaces, used for production of silicomanganese, engineering-and-economical performance are analyzed. Questions of data of the electric, thermal and technological modes of the furnace functioning collecting and processing for use in operation of an advanced control system of the furnace providing increase in technical and economic efficiency of technological process and an adaptability to quality of burden stock are considered.

1. Introduction

This paper discusses the issues of improving engineering-and-economical performance of ore-thermal electric furnaces, used for producing ferroalloys, in particular, of silicomanganese. The article is based on the results of theoretical studies and practical works performed with the aim of improving engineering-and-economical performance in one of the furnaces for producing silicomanganese (with a power of 9 MVA) at "Kuznetskie Ferrosplavy" Company (Novokuznetsk city, Russia).

Need of researches is in the field connected with the fact that the modes of operation of furnaces which have developed in practice often aren't optimum from the point of view of engineering-and-economical performance, especially at the changing quality of burdening materials (ore).

Has the following reserves of improving engineering-and-economical performance of ore-thermal furnaces:

- reduction of electrical losses in current lead and increase in electrical efficiency;
- reduction of heat losses through the liner and the exhaust gases and increase in thermal efficiency;

- reduction in compliance of the electric modes of operation of furnaces to the changing technological conditions and increase in technological efficiency.

The used approaches are applicable not only to furnaces for producing silicomanganese, but also to ore-thermal furnaces for other materials.

2. Experimental investigations on the furnace

To identify opportunities to improve engineering-and-economical performance, experimental investigation on existing furnace, objectives which are:

- Measurement of electrical parameters (for each phase separately) on the sides of high voltage and low voltage, the determination of resistance and reactance of current lead and molten bath in different periods of the technological process of melting, the calculation of the electrical characteristics of the furnace based on the obtained data.
- Measurement of the temperature field of the furnace casing by the thermal imager for gauge of heat losses based on measurement.
- Measurement of temperature and air flow rate in each channel of the cooling hearth to assess the cooling efficiency and gauge of heat losses.
- Measurement of temperature of busbars at different points from the terminals of the transformer to the flexible part in the operating mode for the evaluation of conditions of current lead.
- Measure temperature of exhaust gases and melt, discharged from the furnace, for gauge of heat losses.
- Spectrum analysis of voltage and current for power quality.
- Formation of the file of electro-technological parameters of furnace operation (for the month), its mathematical processing and analysis to determine a rational regime of melting.

Before carrying out experimental investigation on the furnace the theoretical calculation of resistances and reactances of all phases of furnace circuit and building of the electrical characteristics of the furnace for operating voltage steps of the transformer by the method of [1] were carried out. This calculation is required for pre-selection of parameters of electric mode of furnace (current, voltage) on power factor ($\cos\varphi$) and electrical efficiency.

Electrical efficiency is defined as

$$\eta_e = \frac{R_b}{R_b + R_{c.l.}} \quad , \tag{1}$$

where R_b– the resistance of the molten bath, $R_{c.l.}$ – the resistance of the current lead. For a given furnace design value of the electrical efficiency of 0.92.

By results of measurements on the furnace dependence of inductive reactance of an furnace circuit of X_{circ} (milliohm) on current of I (kA) is received. For the considered furnace this dependence is approximated by a power function

$$X_{circ} = 2.263 I^{-0.17} \tag{2}$$

Taking into account this dependence electrical characteristics of the furnace are calculated further.

The operating electric mode of the considered furnace is characterized by the following parameters: the secondary voltage of the transformer of 139.5 V, current of 33.0 kA, apparent power of the furnace is 8 MBA on average, average power factor ($\cos\varphi$) of 0.885. This mode practically corresponds to the parameters chosen on $\cos\varphi$ and electrical efficiency, but is reached only at the expense of a 12 percent overload of the used transformer (current of 33.0 kA at nominal current of 29.4 kA).

Fluctuations of voltage on primary side of the transformer (6 kV) by results of measurements on the operating furnace are small and are in limits of 2% that conforms to qualifying standards.

In the analysis of a range of current existence of significant harmonious components with numbers 2, 3, 5, 7, 23, 25, 35, 37, 47 and 49 is revealed, at the same time values of the 23rd and 25th harmonious components (1,7% and 1,3% respectively) exceed qualities of the electric power, admissible according to the Russian state standards. These harmonious components can exert bad impact on the condenser installations installed in distributing devices of 6 kV and on cable lines of 6 kV, reducing reliability of its work.

Thermal measurements on a casing of the electric furnace were carried out for the purpose of an assessment of heat losses through side liner, and also distributions of these losses on the area of a casing. Measurements are carried out with use of the *Testo 880* thermal imager. Temperatures by sectors into which the surface of a casing is divided by vertical stiffening ribs and horizontal stiffening rings have been taken.

Besides, temperature and speeds of an air stream in channels of compulsory air cooling of a hearth are taken. According to these measurements and thermovision measurements on a casing heat losses through liner of the furnace are calculated (for the considered furnace heat losses are 384 kW: 166 kW through a side surface and 218 kW through a hearth).

Results of thermal measurements can be used for an energy audit, and also for modification of a liner design for the purpose of increase in thermal efficiency at modernization of the furnace.

3. Electrical characteristics on steps of voltage of the furnace transformer

Under electrical characteristics (Figures 1, 2) understand dependences of the active power consumed by the furnace and the useful power (the power which is allocated in a bath) on current for each step of voltage of the furnace transformer. Each line on graphs in Figures 1, 2 corresponds to the step of voltage.

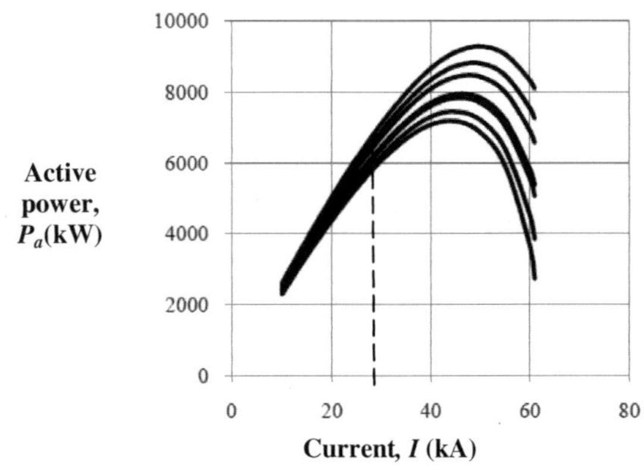

Figure 1. Electrical characteristics of the ore-heating furnace (active power).

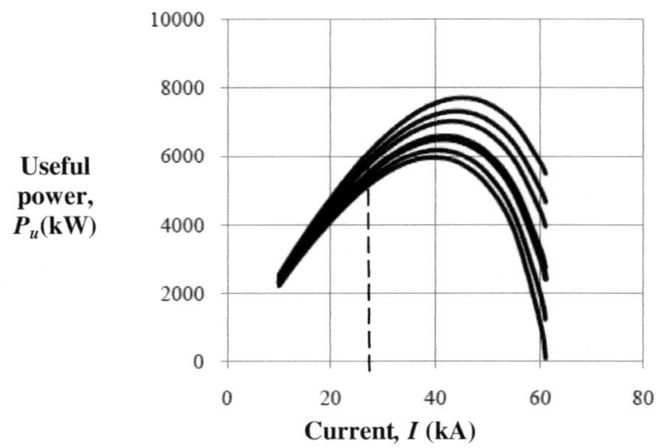

Figure 2. Electrical characteristics of the ore-heating furnace (useful power).

Electrical characteristics are under calculation taking into account the measured value of resistance of a current lead and dependence of inductive reactance of an furnace circuit on current (2).

The analysis of electrical characteristics shows that maxima of active P_a and useful P_u powers don't coincide (the maximum of useful power comes at smaller value of current) and as the course of processes in a molten bath of the furnace is defined by useful power, its maximum should be supported at regulation of the electric mode.

In the considered furnace at the 11th, 13th, 15th and 17th steps of voltage follows from electric characteristics of the considered furnace that the maximum of useful power at these steps is reached at values of current of 52, 49, 46.5 and 43.5 kA (useful power respectively 8330, 7550, 6930 and 6310 kW), however the possibility of achievement of such capacities is limited to admissible current of the transformer of 28 kA (dashed lines in Figures 1, 2) therefore real values of a maximum of useful power at these steps of voltage make 6200, 5840, 5540 and 5220 kW.

It is possible to draw a conclusion that the available furnace transformer doesn't correspond to parameters of this furnace and its current lead, doesn't provide a supply of necessary power with the necessary ratio of voltage and current and demands replacement by more powerful transformer with calculated parameters.

4. Melting electro-technological mode model

High engineering-and-economical performance of operation of the furnace can be provided with only correctly picked up electro-technological operating mode. The choice of a rational electro-technological operating mode of the ore-thermal electric furnace, unlike the electric mode without connection with technology, represents a multiple-factor task in which output indicators (furnace productivity, a specific power consumption, extraction of the leading element, etc.) depend on a large number of the influencing factors which treat the characteristic of the furnace charge (the chemical composition, the size of fraction, humidity on each of charge components), the electric mode, including deepening of electrodes in a molten bath. At the same time even the skilled founder isn't able to support constantly manually the optimum electro-technological mode, especially at frequent changes of conditions during melting.

Maintenance of the optimum electro-technological mode is possible only with use of the mathematical model connecting input parameters of technological process in ore-thermal furnace with output parameters.

It is almost impossible to construct such model by analytical methods as the task of the mathematical description of process is multiple-factor, and character of many factors is probabilistic. Therefore connection between input and output parameters can be established most authentically by collection of information directly on the operating furnace and the subsequent its mathematical processing (the regression analysis) with use of special software.

The advanced control system of ore-thermal electric furnace uses such model.

Formation of the file of input data for creation of regression model consists of several stages:

- Formation the initial file of parameters of the electric mode of the furnace with a 5-minute interval of data recording during from beginning to end each melting, association of the file with data on the analysis and a consumption of raw materials, daily and operational cards of the electric furnace. Sources of information are furnace control systems data, data of registration of electric parameters in real time by the programmable analyzer of energy consumption of the *AR-5* series, and also documentation which is constantly maintained on the furnace by operational personnel.
- An exception of the data which are misses from the file (rejection).
- A filtration (to exclude fluctuations) and averaging of data of the electric mode on necessary time intervals with the purpose to track and determine consistent patterns of their change during melting.
- A filtration and averaging of electro-technological parameters on time intervals lasting one melting for the purpose of determination of the rational furnace operating mode.

Processing of experimental data is based on engineering algorithms.

As a result of processing and the analysis of the file of the data melting processes decide on the most rational electric indicators and engineering-and-economical performance. The selected melting processes are grouped in the range of change of the size ctgφ from 4 to 5.5 (φ is a phase angle between voltage and current). Angle φ cotangent value is directly connected with deepening of electrodes, with the characteristic of composition of the furnace charge, with engineering-and-economical performance of the furnace operation, etc.

Examples initial and smoothed (by a filtration and averaging) dependences of electric parameters (active power and current) on time during melting are presented in Figures 3, 4.

As a first approximation each melting can be broken into three time intervals which correspond to an initial stage of melting after melt discharge (a smooth growth in power to the maximum value with growth in resistance and decrease in reactance of a bath), to the most long period of the stable electric mode and the final period before melt discharge. Electrical characteristics on intervals (in dynamics during melting) allow to receive the optimum design mode of this melting and to compare it with the actual mode.

IOP Conf. Series: Materials Science and Engineering **313** (2018) 012012 doi:10.1088/1757-899X/313/1/012012

The choice of the optimum mode in case of fluctuations of quality of furnace charge and technology of melting demands continuous processing of information arriving from the furnace in real time with updating of data file.

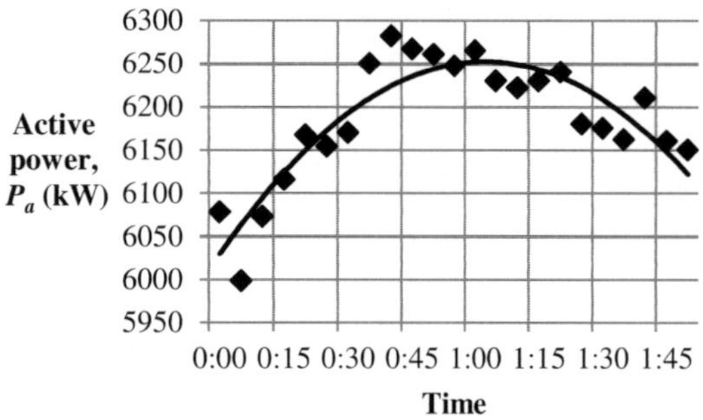

Figure 3. Change of active power during melting.

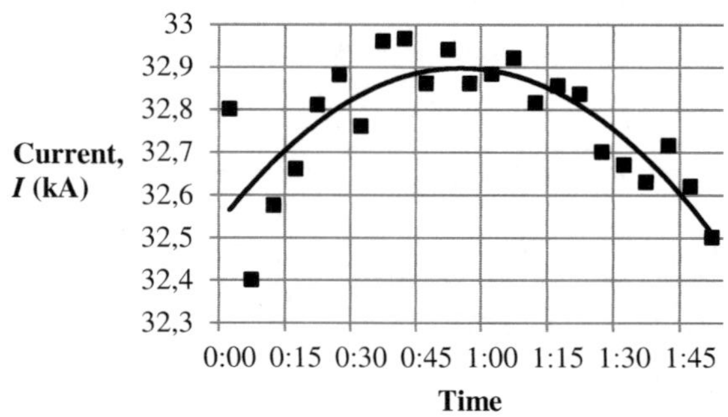

Figure 4. Change of current (average value on phases) during melting.

The data file received on the operating furnace is presented in the form of a matrix which lines contain values of entrance parameters factors (usually 8–12 most technologically significant) and output indicators (usually 3–4). This matrix with a time interval in one melting is processed by means of specially developed software of the regression analysis. Receiving the equation of regression connecting values of each of output indicators y_j with factors x_1, x_2, ... , x_n is result of processing. The equation of regression has an appearance of a square polynomial

$$y = b_0 + b_1x_1 + b_2x_2 + \cdots + b_nx_n + b_{1\,2}x_1\,x_2 + \cdots + b_{n-1\,n}x_{n-1}\,x_n + b_{1\,1}x_1^2 + b_{2\,2}x_2^2 + + \cdots + b_{nn}x_n^2 \quad (3)$$

Content of manganese ore, reducer (coal), dolomite and slag in furnace charge, the active power, operating voltage, electrode current can be examples of factors x_i, examples of output indicators y_j are productivity of the furnace (t/h), and a specific expense of the electric power (kWh/t). At the choice of the rational electro-technological mode it is expedient to use the integrated indicator of efficiency of process *IIE* considering (with various weight coefficients k_j received by an expert assessment) various output engineering-and-economical performance

$$IIE = k_1y_1 + k_2y_2 + \cdots + k_my_m. \quad (4)$$

Use of the software of the regression analysis demands installation on the furnace of the registering measuring devices (for registration of electric and power parameters) and records of entrance technological parameters and output operational performance in the form (manual input) coordinated in advance for further automatic processing of data file.

Processing of the prepared file of the experimental data obtained on the studied furnace with use of the software of the regression analysis is carried out in the following order:

1. Coefficients of the polynomial of regression (3) for an integrated engineering-and-economical performance (indicator of efficiency) are determined.
2. Little significant input parameters are excluded.
3. The extremum of the response surface described (3), that is combination of values of factors at which the best value of a engineering-and-economical performance is reached is found. The corresponding value of current (at the known characteristics of furnace charge) is used by the furnace current regulator as a set point, value of active power – for the choice of a step of voltage (according to electrical characteristics). Also on the basis of an extremum of a response surface it is possible to carry out correction of composition of furnace charge for achievement of the best engineering-and-economical performance.

The software of data processing is integrated into an advanced control system of the furnace, the file of input data is formed in the automatic mode. The calculated values of current, power, furnace charge parameters corresponding to the optimum electro-technological mode can be given the program in the form of advice to the operator. At the subsequent stages of introduction and development of advanced control system transition to operation in the supervisory mode with automatic change of set points of current and voltage according to the calculated optimum mode is possible.

5. Conclusion

Theoretical and experimental investigation have found opportunities to improve engineering-and-economical performance of ore-thermal furnaces, used for producing ferroalloys, including, connected with reduction in compliance of the electric modes of operation of furnaces (current, step of voltage) to the changing technological conditions and increase in technological efficiency. Maintenance of the optimum electro-technological mode is possible only with use of the mathematical model connecting input parameters of technological process in ore-thermal furnace (current, a step of voltage, the characteristic of furnace charge, etc.) with output parameters. Such model makes by collection of information directly on the operating furnace and the subsequent its mathematical processing (the regression analysis) with use of special software, and information arriving from the furnace continuously is processed in real time with updating of data file.

The special software integrated into an advanced furnace control system and realizing the offered technique of collecting and data processing in the automatic mode allows to pick up the electric mode of melting at which the best value of an integrated indicator of efficiency of process *IIE* (which considers the furnace productivity, a specific expense of the electric power, extraction of the leading element, indicators of the electric power quality, etc.) is reached. Calculation of correction of furnace charge composition for achievement of the best engineering-and-economical performance is also possible.

References

[1] Dantsis Ya B 2007 *Secondary current leads and electrical parameters of arc electric furnaces* (Moscow: Metallurgy) 320 p

[2] Kondrashov V P, Kolyvanov S Yu, Lykov A G, Pogrebisskiy M Ya, Saprykin A I and Savalyk N A 2010 Rational operating conditions for ore-heating electrofurnaces *Steel in Translation* Volume **40**, Issue **2**, pp 145-152

[3] Kondrashov V, Pogrebisskiy M and Salmanova E 2015 Development of ways of increase of ore-heating electric furnaces for production of ferroalloys efficiency *Proc. Int. Conf. AMTEE'15 – Advanced Methods of the Theory of Electrical Engineering* (Pilsen: University of West Bohemia) p I-5

[4] Mraz J 2011 The interpretation of electrical measurements of submerged arc-resistance furnace *Proc. Int. Conf. AMTEE'11 – Tenth International Conference on Advanced Methods in the Theory of Electrical Engineering* (Pilsen: University of West Bohemia) pp VI-5-VI-6

[5] Machulec B 2011 Equilibrium model of the ferrosilicon melting process in the submerged arc-resistance furnace *Proc. Int. Conf. AMTEE'11 – Tenth International Conference on Advanced Methods in the Theory of Electrical Engineering* (Pilsen: University of West Bohemia) pp VIII-3-VIII-4

Developing the mechatronics and robotics at Nizhny Tagil Technological Institute of Ural Federal University

V V Goman, S A Fedoreev

Ural Federal University, 622031, Nizhny Tagil, Krasnogvardeyskaya str, 59, Russia, vvg_electro@hotmail.com

Abstracts. This report concerns the development trends of education in the field of the Mechatronics and Robotics at Nizhny Tagil Technological Institute (branch of Ural Federal University). The paper considers new teaching technologies, experience in upgrade of the laboratory facilities and some results of development Mechatronics and Robotics educational courses.

1. Introduction

In 2017, Nizhny Tagil Technological Institute (NTI), a branch of Ural Federal University (UrFU), graduated its first bachelors in Mechatronics and Robotics. This new and complicated professional qualification has been mastered due to the cooperated effort of multiple chairs, due to advanced training of the professors, new teaching technologies, and development and upgrade of the laboratory facilities.

2. Development of laboratory facilities

At the outset, the institute was equipped with a number of training laboratories that enabled teaching of some subjects of this discipline. However, the laboratory equipment was partly obsolete and dispersed among different buildings. It needed to be replenished and consolidated by a single concept of mechatronics.

To this end, the hydraulics and pneumatics laboratory [1] - the most advanced and spacious one - got new electric drives, an automation system, and an industrial data transmission network. This network (powered by Siemens equipment) united the available laboratory benches of different operation principles into a single mechatronic system.

The laboratory has contributed to the training of students and retrainees and thus improved the quality of training in industrial automation and mechatronics (that included programming of Siemens controllers, industrial information networks, visualization system, electric and hydraulic drives etc.). Besides students specialized in Mechatronics and Robotics, the laboratory has been of great use for the students of Design and Technology Supports of Engineering Works; students of numerous technical professions have been studying Electronics, Automation of Production Processes, Control Systems, Electric Drives and other disciplines there. The laboratory is being used in such projects as New Staff for Defense Industry Complex, and training in the System Engineering course for the Higher Engineering School of UrFU.

For this reason, the laboratory now features workbenches with electric, hydraulic and pneumatic drives, automation equipment (programmable logical controllers - PLC and accessories), personal computers, human-machine interface (HMI) devices, and integrated industrial data transmission networks (Profibus and Profinet).

IOP Conf. Series: Materials Science and Engineering **313** (2018) 012013 doi:10.1088/1757-899X/313/1/012013

Figure 1. S7-300 PLC-based workbench

Figure 2. S7-1200 PLC-based workbench

Figure 1 shows a typical PLC-based training workbench. This workbench includes the Siemens PLC S7-300 [2]. S7-300 is widely used in industry for medium complexity automation tasks. In figure 2, you see a S7-1200 PLC-based laboratory workbench for training in pneumatic drive and pneumatic automatics. S7-1200 [3] is PLC for small and medium automation systems. Figure 3 gives training laboratory workbench based on Sinamics electric drives.

All the PLCs, PCs and operator panels are united with a single data transmission network to simplify the interaction between students and professors and enable study of various industrial data transmission networks and fieldbus networks.

The laboratory also features:

1. PLC-equipped workbenches imitating workshop mechanisms (turn table, distribution system, pump station, electric heating station, positioning system);
2. educational robotics construction kits based on Arduino microcontrollers, used for practical trainings and lessons of the robotics section for school students;
3. a CNC milling machine by Purelogic, 3D-printer;
4. a virtual hands-on course allowing remote operation of a real PLC with virtual controlled objects (figure 4) [4, 5, 6];
5. a software package.

Figure 3. A workbench based on electric drives by Sinamics

IOP Conf. Series: Materials Science and Engineering **313** (2018) 012013 doi:10.1088/1757-899X/313/1/012013

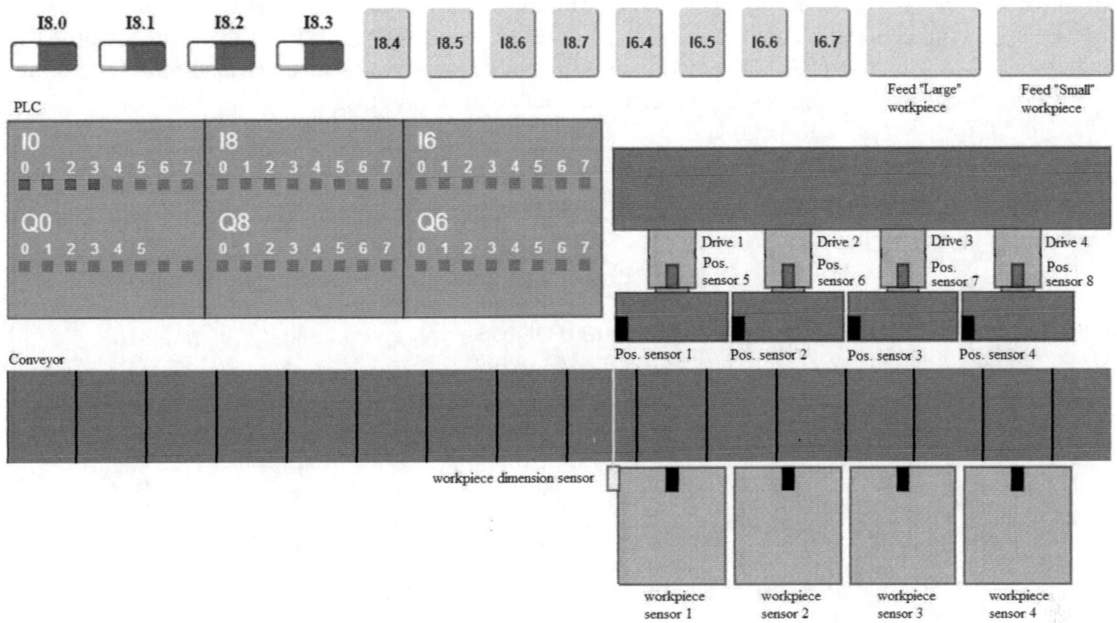

Figure 4. Virtual lab workbench: Sorting Conveyer

Also we can use the equipment of our industrial partners (EVRAZ and Uralvagonzavod) - Festo MPS (modular production system) and MTLR (production line with turning and milling CNC-machines and robot on linear axis) systems.

3. Education technologies

The courses apply the world's best practices of engineering education including, first of all, CDIO initiative (Conceive – Design – Implement – Operate) [7] by Massachusetts Institute of Technology (also referred to as Project-Based Learning). Today, this project unites over 100 universities all over the world, including a number of Russian institutes, with Ural Federal University among them.

Project practices for bachelors are moderated by project engineers – postgraduate students chosen from the most talented bachelors (simultaneously, they are being involved in developing real projects).

The general supervision and goal-setting are provided by profession supervisors: chair professors, educational program leaders. When necessary, the courses apply to competent experts, consultants, and plant employees. Different professions are consolidated by a multidisciplinary Students' Design Bureau, which residents being guided by their supervisors and experts are able to work through a certain idea, provide project docs and calculations to finally implement it as a real product or technology (figure 5).

As new projects keep evolving into real products, we are intended to take part in programs of Strategic Initiative Agency (National Technology Initiative), Foundation of Innovation Promotion (UMNIK, START, and Commercialization programs). The goal is to establish small innovative businesses (start-ups, spin-off companies with participation of UrFU).

According to this strategy, start-ups, along with big corporate customers, are to generate tasks for students' project activities.

Looking at the first graduation of the Mechatronics and Robotics qualification, we can confidently state that the efforts have yielded some results. For most graduates, their final qualification work is intermediate stage of work over their project. Three project concepts have been supported by the Foundation of Innovation Promotion (UMNIK and START programs). They are a mechanic therapy tool to recover limbs after injuries and treat cerebral palsy, an innovative machine for protective coating of electric contacts (dozen times more effective than manual application); a robotic machine for electrosparking application of coating in hard-to-reach locations such as inside of pipes. Some final qualification works 2017 has developed training laboratory workbenches showing positioning systems, electric drive machine-tools, and electric-heat control systems.

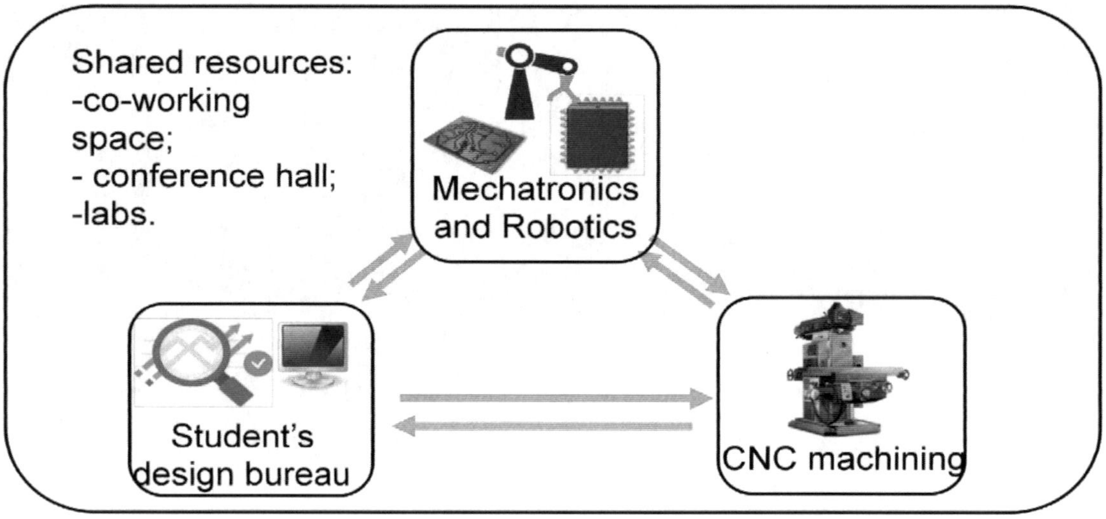

Figure 5. Interactions within Project Learning

4.Plans and prospects

We plan to focus on such disciplines as artificial intelligence, machine learning, industrial robotics, and system engineering. To train practical skills, we are training student teams to participate in WorldSkills interuniversity championships. A license for master degree education program has been obtained.

References

[1] ZAO Ecoinvent. http://www.ecoinvent.ru/products/ (accessed on 25 January 2018).

[2] SIMATIC S7-300. https://w3.siemens.com/mcms/programmable-logic-controller/en/advanced-controller/s7-300/pages/default.aspx (accessed on 25 January 2018).

[3] SIMATIC S7-1200 – Be flexible thanks to networking possibilities. https://www.siemens.com/global/en/home/products/automation/systems/industrial/plc/s7-1200.html (accessed on 25 January 2018).

[4] Shyr, W.-J. 2010 Enhancement of PLC programming learning based on a virtual laboratory. *World Transactions on Engineering and Technology Education*, **8 (2)**, p 196-202.

[5] Perdukova, D. Fedor P. 2013 Virtual laboratory for the study of technological process automation. *International Journal of Engineering Education*. **29 (1)**, p 230-238.

[6] Shyr, W.-J. Su, T.-J Lin, C.-M. 2013 Development of remote monitoring and a control system based on PLC and web access for learning mechatronics. *International Journal of Advanced Robotic Systems*, **10 (5).** DOI: 10.5772/55363

[7] Welcome to CDIO. http://www.cdio.org/ (accessed on 25 January 2018).

IOP Conf. Series: Materials Science and Engineering **313** (2018) 012014 doi:10.1088/1757-899X/313/1/012014

Development of the self-learning machine for creating models of microprocessor of single-phase earth fault protection devices in networks with isolated neutral voltage above 1000 V

B B Utegulov, A B Utegulov, S Meiramova

Energy Department, S. Seifullin Kazakh Agro-Technical University, 62 Pobeda avenue, Astana, Republic of Kazakhstan

bolatu@mail.ru

Abstract. The paper proposes the development of a self-learning machine for creating models of microprocessor-based single-phase ground fault protection devices in networks with an isolated neutral voltage higher than 1000 V. Development of a self-learning machine for creating models of microprocessor-based single-phase earth fault protection devices in networks with an isolated neutral voltage higher than 1000 V. allows to effectively implement mathematical models of automatic change of protection settings. Single-phase earth fault protection devices.

1. Introduction

One of the most important issues in the electric power industry today is the problem of improving the reliability of power supply and electrical safety in the operation of electrical installations. This condition is associated with the physical and moral aging of most of the equipment, as well as a low level of technical perfection of relay protection and automation.

The prevailing type of damage in 6-10 kV electric networks is single-phase earth faults, which constitute about 75% of the total number of injuries. The duration of the action of single-phase earth fault leads to phase-to-phase and multi-place short circuits.

In many cases, the cause of the closure and further development of the accident is the wear and tear of the electrical insulation of the network. Conditions for electrical safety and reliability of energy supply are largely determined by the state of insulation, its resistance and capacity relative to the ground. Therefore, timely detection and elimination of defects accompanied by a change in the active and capacitive components of conductivity of insulation will prevent most of the damage.

Current methods for determining insulation parameters for measurements require the involvement of personnel to connect additional capacitive conductivity and additional active conductivity to the electrical network, visual removal of voltage modules, manual calculation of insulation parameters, which leads to a reduction in the level of electrical safety of personnel engaged in determining the current single-phase earth fault.

Modern directional protection against single-phase earth faults has a low sensitivity and a large number of false positives. In this case, disconnections are often of a group nature, and it is difficult to find a damaged feeder in the distribution network, as a result of which a single-phase ground fault can become a more dangerous type of fault - two-phase and three-phase. The main drawback of existing protection against single-phase earth faults is the difficulty in selecting the set-point of the pick-up current, since during the operation of the network the number of simultaneously operating connections changes, and consequently the current of the single-phase earth fault also changes. The existing protections do not take into account the position of the automatic switchover sectionalizer, since during sectioning the value of the real current of a single-phase earth fault will also depend on the parameters of the outgoing lines of the connected section and the number of equipment connected to them. Thus, fixed setting of the single-phase earth fault current does not ensure the selectivity of the protection operation when the number of connections in the load node is switched on.

On the basis of the foregoing, it follows that the task of improving the reliability of the internal power supply system of industrial enterprises and ensuring the growth of the level of electrical safety in the operation of electrical installations is the development of microprocessor-based methods and means of automatic determination of insulation parameters and centralized protection against single-phase earth faults with automatic change of the current set point and with a sectional automatic switchover switch in electric networks with voltage 6 - 10 kV power industrial, mining and rural enterprises, is relevant and has important scientific and practical significance [3, 5 - 10].

2. The main points of the self-learning machine when developing models of microprocessor-based single-phase earth fault protection devices in networks with isolated neutral voltage above 1000 V

At present, the introduction of more sophisticated selective protection devices for single-phase earth fault protection based on the use of one of the known methods of operation of these devices can only to a certain extent improve the situation in 6-10 kV networks in enterprises.

The existing methods for determining the regime of the current-preceding single-phase earth fault are not unique. This is due to the fact that the magnitude of the earth fault current and its nature depends on the network parameters, the neutral operation mode, the transient resistance at the fault location and the number of connected voltage transformers.

To date, the most promising technical means of automation are microprocessor means. Therefore, it seems expedient to develop automatic protection systems against single-phase earth faults based on microprocessor means. Therefore, more significant changes can be achieved when developing and implementing microprocessor devices, including multifunctional ones, whose output information would be the result of analyzing the single-phase earth fault process for several factors.

The mathematical model of microprocessor devices, as well as any other digital devices, are finite state machine. At the same time, modeling of such systems consists in the synthesis of finite state machine, which is in an objective correspondence with the real system of protection. However, under the conditions of constantly changing random external influences (which is typical for real open systems), static deterministic machine cannot serve as an adequate model of real objects. The most optimal model of such objects can be self-learning stochastic machine. The main functions of such kind of machines are the accumulation of information about the object, the analysis of the obtained data and the correction of the algorithm of its functioning based on the results of data analysis. Given the random nature and the degree of reliability of the accumulated data, the Monte Carlo method can be used for their analysis, which is widely used to solve certain statistical problems that cannot be analyzed analytically [1, 2, 4].

However, before talking about self-learning machines, it is necessary to understand what the learning process is. Perhaps the following definition: "Learning is a change in behavior or functioning based on information about the outside world. At the same time, the change must occur in the direction of improving behavior or action in the desired direction. " An essential component of learning is the memorization (accumulation) of information.

The concept of "training of system (machine)" is defined as follows:

The training of the system consists in accordance with previous successes or failures (experience), it improves the internal model of the external world.

In this connection, two types of self-learning systems (machines) are possible.

IOP Conf. Series: Materials Science and Engineering **313** (2018) 012014 doi:10.1088/1757-899X/313/1/012014

2.1. Self-learning automatic system

A self-learning automatic system is a learning machine or a self-adaptive system whose control algorithm changes in accordance with the evaluation of the control results so that over time it improves its characteristics and the quality of functioning. In general, the design and construction of technical systems is possible only on the basis of the initial a priori information about the nature of processes occurring in the system and the conditions accompanying the operation of the system and exerting a disturbing influence on it. When there is complete initial a priori information, it is possible to accurately determine such values of the characteristics of the system being designed, which ensure a given quality of its functioning; In this case, there is no need for its training. In the absence of complete initial information, the only way to create a system with a given quality of functioning is to use the training principle in its development.

Training is the process of repeated influences on the system and the correction of its reactions to these impacts. External adjustment, or, as it is also called, "encouragement" and "punishment", is carried out by the "teacher" who knows the desired reaction to certain influences. "Teacher" can be either a person - an operator, or a machine. It is on the basis of the processing of the control (a posteriori) information that the missing initial information is replenished. If training is carried out without an external training device, then such a system is called self-learning.

Training is carried out by using algorithms, depending on whether the learning automatic system is discrete or continuous, and represent a system of stochastic difference or stochastic differential equations. Training algorithms are implemented by means of computer technology - digital or analog computers (in particular, by electro integrators) or, finally, by hybrid computer systems. As learning, the self-learning automatic system accumulates experience, on the basis of which the required system response to external influences is gradually developed; The self-learning automatic system is an asymptotically optimal system, because the optimal response of the system to external disturbances is not achieved immediately, but over time, as a result of training.

2.2. Self-learning system

The self-learning system is a self-adjusting system whose functioning algorithm is developed and improved in the process of self-learning. This process is reduced to "samples" and "mistakes". The system performs trial changes to the algorithm and simultaneously monitors the results of these changes. If the results are favorable in terms of management objectives, the changes continue in the same direction until the best results are achieved or until the management process deteriorates. Self-learning systems include as a private kind of search self-tuning systems.

Any self-learning machine has the ability to model. The system has the ability to model if it defines reactions to its various (models) actions with the help of the conducted external world model, before taking any action, directed to the outside and performs only those actions, which cause the desired reaction of the external world.

To implement the above, we present the block diagram of a model of a microprocessor-based single-phase earth fault protection system in networks with an isolated neutral voltage higher than 1000 V which can be represented as follows on figure 1:

IOP Conf. Series: Materials Science and Engineering **313** (2018) 012014 doi:10.1088/1757-899X/313/1/012014

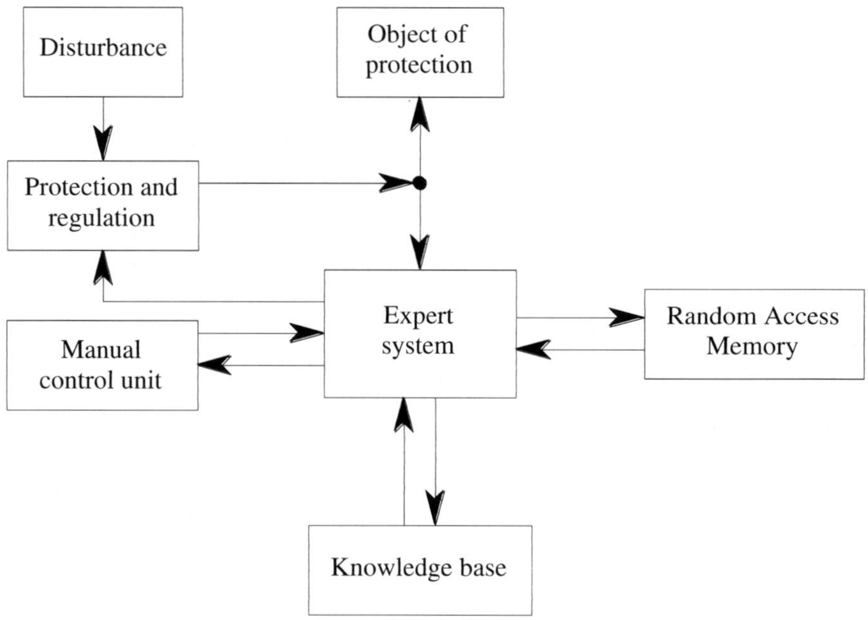

Figure 1. Block diagram of a model of a microprocessor based single-phase earth fault protection system.

The whole Model is presented in the form of the following theses:

- The model of a microprocessor system is a finite abstract self-learning machine with a nondeterministic input alphabet.
- Data on the object are divided into two types:
    - current - "short-lived" accumulated in the operative memory and subjected to continuous analysis;
    - accumulated (knowledge) - "long-lived" accumulated as a result of analysis of current data, stored in permanent memory and periodically analyzed.
- The duration of the time interval between two successive procedures for the removal of information about the protection object is directly dependent on the accumulated knowledge (the results of the analysis of current data);
- Based on the analysis of all data about the object, the model changes its logical structure (the functioning algorithm).
- Mandatory presence of a model (variable or unchangeable) of an ideal protection object.

In accordance with the block diagram of the microprocessor protection system model from single-phase earth fault, it is possible to build automatic monitoring systems for capacitive fault currents in 6-10 kV networks of power, industrial, mining and rural enterprises with the following advantages:

- simplicity of construction of control units providing speed and sensitivity in single-phase ground-fault mode;
- multifunction, which allows you to combine several different protections in one device, the ability to register processes and events, as well as the location of the damage;
- easy installation and operation, as well as a significant reduction in the time for testing terminals using special tools and data processing programs and adjustment of microprocessor devices;
- wide diagnostic system with reporting of a malfunction;
- a variety of types of human-machine communication that brings the microprocessor device closer to the user, regardless of its location;

- the creation of flexible structures that enable the integration of microprocessor-based protection and automation devices in the Automated Process Control System.

3. Conclusion

The development of a self-learning machine in creating models of microprocessor protection devices from single-phase earth faults in networks with an isolated neutral voltage higher than 1000 V makes it possible to effectively implement mathematical models of automatic changes in protection settings.

Development of microprocessor-based methods and means of automatic and centralized protection against single-phase earth faults with automatic change of the electric current setting and with a sectional switch of automatic reserve input in 6-10 kV of electric networks of power, in industrial, mining and rural enterprises, provide increased reliability of the internal power supply system of the enterprise and ensuring the growth of the level of electrical safety in the operation of electrical installations.

References

[1] Baranov S I 1979 *Synthesis of microprogram machines* (Leningrad: Energiya) p 219

[2] Kudryavtsev V B, Aleshin S V and Podkolzin A S 1985 *Introduction to automata theory* (Moscow: Nauka) p 320

[3] Utegulov B B, Utegulov A B , Begentaev M, Zhumazhanov S and Zhakipov N 2011 *Method for determining parameters of isolation network voltage up to 1000 V in mining enterprises (Source of the Document Proc. of the IASTED Int. Conf. on Power and Energy Systems and Applications, PESA)* Proc. 756-028 (Pittsburgh: ACTA Press)

[4] Lupal A M 2000 *Automata theory: Tutorial* (Saint Petersburg: Saint Petersburg State University of Aerospace Instrumentation) p 119

[5] Utegulov B B, Utegulov A B and Zhankuanyshev M 2015 Increasing the reliability of power supply and the safety at operation of electroreceivers in the networks of 6-10 kV *Journal of scientific research and development* **2** special issue 7 (September)

[6] Utegulov B B 2016 *The method of determining the insulation parameters in three-phase electrical networks with isolated neutral with voltages up to and above 1000 V (Int. Conf. on New Energy and Future Energy System)* vol 40 (Beijing: IOP Conference Series, Earth and Environmental Science)

[7] Utegulov B B 2016 *Development of mathematical models of digital protection devices from a single phase earth faults in networks with voltage 6 – 10 kV (2nd Int. Con. on Industrial Engineering, Applications and Manufacturing – ICIEAM)* DOI: 10.1109/ICIEAM.2016.7910979 (Chelyabinsk: IEEE Xplore Conf. Proc.)

[8] Utegulov B B, Utegulov A B and Uakhitova A B 2016 Increasing the efficiency of protective devices for mining machines *Journal of Mining Science (article on russian language)* **2** 102–108

[9] Utegulov B B, Utegulov A B and Uakhitova A B 2016 Development of Method to Improve Efficiency of Residual Current Device under 1000 V on Excavators of Mining Enterprises *Journal of Mining Science* **52** 325–331

[10] Utegulov B B, Utegulov A B and Uakhitova A B 2017 About the method of measuring the total conductivity of insulation in a low-voltage network with isolated neutral at mining enterprises *Journal of Mining Science (article on russian language)* **2** 77–85

Analysis of the error of the developed method of determination the active conductivity reducing the insulation level between one phase of the network and ground, and insulation parameters in a non-symmetric network with isolated neutral with voltage above 1000 V

B B Utegulov

Energy Department, S. Seifullin Kazakh Agro-Technical University, 62 Pobeda avenue, Astana, Republic of Kazakhstan

bolatu@mail.ru

Abstract. In the work the study of the developed method was carried out for reliability by analyzing the error in indirect determination of the insulation parameters in an asymmetric network with an isolated neutral voltage above 1000 V. The conducted studies of the random relative mean square errors show that the accuracy of indirect measurements in the developed method can be effectively regulated not only by selecting a capacitive additional conductivity, which are connected between phases of the electrical network and the ground, but also by the selection of measuring instruments according to the accuracy class. When choosing meters with accuracy class of 0.5 with the correct selection of capacitive additional conductivity that are connected between the phases of the electrical network and the ground, the errors in measuring the insulation parameters will not exceed 10%.

1. Introduction

The decrease in the level of insulation resistance of the phases of the electrical network relative to the ground creates the occurrence of emergency operating modes of electrical installations, which may result in electric shock to people.

In order to eliminate the electric shock to people, it is necessary to provide a high level of insulation resistance in a network with an isolated neutral voltage of more than 1000 V by carrying out activities related to systematic and effective monitoring of its condition, which is one of the main directions for providing electrical safety in a 6 kV auxiliaries of power plants.

The practice of operating electric grids with a voltage of 6 kV at thermal power plants shows the absence of a method for measuring insulation resistance at the enterprise. And if it is available, the insulation resistance measurements are made, as a rule, extremely irregularly and, moreover, with large errors. The most widely used method was the measurement of insulation resistance by using a measuring device - megger.

Unsatisfactory installation of preventive inspections and repairs of operating electrical installations, as well as worsening of the insulation of networks and electrical equipment, create the danger of occurrence of emergency operating conditions, which may result in electric shock, equipment damage and long down-time of high-performance mechanisms. In the process of operating the electric power plant, maintenance personnel rely mainly on protection against single-phase earth faults, which may be faulty.

In the organization of works taken at thermal power plants, time is not always allocated for execution and implementation in full the repair, commissioning and preventive testing of electrical installations [1-11]. Meggers used to measure the resistance of phases of electrical installations relative to ground cannot be considered a reliable means of prevention.

The other side of this issue is the lack of clear recommendations on the method of measuring the insulation resistance in a 6 kV network of auxiliary needs of power plants in accordance with the requirements of the Uniform Rules of Safety. Numerous proposals on methods for measuring the insulation parameters of electrical installations with voltages above 1000 V have hitherto not led to the creation of a universal control method suitable for both research and practical purposes.

In connection with the foregoing, the urgency of the task of further improving the means for continuous and periodic monitoring of the insulation parameters of a network with voltages above 1000 V of industrial and mining enterprises in the process of their operation continues to be relevant. Operation of a network with a voltage above 1000 V in industrial and mining enterprises with low insulation resistance is very dangerous, and therefore the issue of controlling insulation resistance and maintaining it at a high level should be given constant attention. style).

2. Analysis of the error of the developed method for determining the insulation parameters in a network with a voltage above 1000 V

In three-phase electrical networks with an isolated neutral voltage above 1000 V, there is a lack of symmetry of phase voltages. Non-symmetry in high voltage networks is formed with three factors: unbalanced load on the phases of the electrical network; incorrect execution of transposition of high-voltage wires; when one phase of the network is broken under load; if any phase relative to the ground is damaged.

One of the important priorities of the developed indirect methods of determining the insulation parameters in a network with an isolated neutral voltage above 1000 V is the accuracy of determining the unknown quantities. In the indirect determination of the isolation parameters in a network with an isolated neutral, strict requirements are put in place to ensure the reliability of the values of the unknown quantities.

Carried out analysis of the error of the developed method for determining the active conductivity, which reduces the isolation level between one of the phases of the network and ground, and of the insulation parameters in an asymmetric network with an isolated neutral voltage above 1000 V, which is based on the measurement of the values of the line voltage and the residual voltage, as well as was carried out the analysis of the angle of lag between the vectors of the line voltage and the residual voltage before and after connecting additional capacitive conductivity between the phases of the network and ground, and determine:

- the conductivity value, which reduces the insulation resistance of one of the phases of the electrical network relative to the earth:

$$g_o = \frac{U_o U_{o1}}{3U_l(U_o\cos\alpha_1 - U_{o1}\cos\alpha)} b_o, \qquad (1)$$

- determination of capacitive conductivity of network isolation:

$$b = \frac{U_{o1} b_o \cos\alpha}{U_o\cos\alpha_1 - U_{o1}\cos\alpha}, \qquad (2)$$

- active conductivity of network isolation:

$$g = \frac{(U_l\sin\alpha - U_o)U_{o1} b_o}{1.73 U_l(U_o\cos\alpha_1 - U_{o1}\cos\alpha)}, \qquad (3)$$

- full conductivity of network isolation:

IOP Conf. Series: Materials Science and Engineering **313** (2018) 012015 doi:10.1088/1757-899X/313/1/012015

$$y = \frac{U_{pho} U_{o1} b_o}{1.73 U_l (U_o \sin\alpha_1 - U_{o1} \sin\alpha)}, \tag{4}$$

where U_l – line voltage of the network;

U_o – zero sequence voltage in case of insulation failure of one of the phases of the network relative to the ground;

$\cos\alpha$ – angle of lag between the vectors of the voltage of the zero sequence and the line voltage;

U_{o1} – zero sequence voltage in case of insulation failure of one of the phases of the network relative to the ground after connecting additional capacitive conductivity between the phases of the network and ground;

$\cos\alpha_1$ – the angle of lag between the zero sequence voltage and line voltage vectors, after connecting additional capacitive conductivity between the phases of the network and ground;

b_o – additional capacitive conductivity, connected between phases of the network and ground.

In order to analyze the error of the developed method for determining the insulation parameters in an asymmetric network, it is necessary to fulfill the replacement conditions U_l - linear voltage to U_{ph} – the phase voltage relative to the ground before the insulation of the network is damaged, since $U_l = 1.73 U_{ph}$ it is also assumed that when unbalance appears, the line voltages remain unchanged. The zero sequence voltage on the open triangle of the voltage transformer is characterized as $3U_o$, with taking into account the above described equations, the parameters describing the insulation will take the form:

- the conductivity value, which reduces the insulation resistance of one of the phases of the electrical network relative to the ground:

$$g_o = \frac{U_o U_{o1}}{U_{ph}(U_o \cos\alpha_1 - U_{o1} \cos\alpha)} b_o, \tag{5}$$

- active conductivity of network isolation:

$$g = \frac{(U_{ph} \sin\alpha - U_o) U_{o1} b_o}{U_{ph}(U_o \cos\alpha_1 - U_{o1} \cos\alpha)}, \tag{6}$$

- full conductivity of network isolation:

$$y = \frac{U_{pho} U_{o1} b_o}{U_{ph}(U_o \sin\alpha_1 - U_{o1} \sin\alpha)}. \tag{7}$$

It should be noted that when determining the insulation parameters in a network with an isolated neutral, based on the measurement of the magnitude of the residual voltage modules and the phase voltage relative to the ground, and the angle of lag between the residual voltage and linear voltage vectors when connecting the capacitive additional conductances between the phases of the network and ground , it becomes necessary to establish the methodological error of the obtained mathematical dependences.

Based on the analysis of error, practical recommendations are developed that ensure:

- normal operation of electric receivers in the production of measurements;

- the satisfactory accuracy of the developed method for determining the insulation parameters in a network with an isolated neutral voltage above 1000 V, with the achievement of safety in the production of works in electrical installations, and also with the provision of simplicity and convenience of measurements.

When analyzing the error of the developed method for determining the insulation parameters of the phases of the electrical network relative to the ground in a network with an isolated neutral voltage above 1000 V, the determined ratio of the relative method error to the error of the measuring devices measuring the magnitude of the voltage modules and the value of the capacitive conductivity is used as additional conductivity.

The error analysis of the developed method for determining the insulation parameters of an asymmetric network with a voltage above 1000 V is made using the basic provisions of the theory of errors and the theoretical foundations of electrical engineering.

The relative rms error for conductivity reduces the insulation resistance between one of the phases of the network and ground in networks with an isolated neutral voltage above 1000 V is determined from the equation (5):

$$g_o = \frac{U_o U_{o1} b_o}{U_{ph}(U_o \sin\alpha_1 - U_{o1}\sin\alpha)},$$

where U_{ph}, U_o, U_{o1}, $\sin\alpha$, $\sin\alpha_1$, b_o are the values that determine the conductivity that reduces the insulation resistance between one of the phases of the network and ground in networks with an isolated neutral with voltage above 1000 V, obtained by direct measurement.

The relative root-mean-square error of the method for determining the conductivity of a resistance-reducing insulation between one of the phases of the network and ground in networks with an isolated neutral voltage higher than 1000 V is determined from the equation [6-9]:

$$\varepsilon_{g_o} = \frac{1}{g_o}\left[\begin{aligned}&\left(\frac{\partial g_o}{\partial U_{ph}}\right)^2(\Delta U_{ph})^2 + \left(\frac{\partial g_o}{\partial U_o}\right)^2(\Delta U_o)^2 + \left(\frac{\partial g_o}{\partial U_{o1}}\right)^2(\Delta U_{o1})^2\\ &+\left(\frac{\partial g_o}{\partial \sin\alpha}\right)^2(\Delta\sin\alpha)^2 + \left(\frac{\partial g_o}{\partial \sin\alpha_1}\right)^2(\Delta\sin\alpha_1)^2 + \left(\frac{\partial g_o}{\partial b_o}\right)^2(\Delta b_o)^2\end{aligned}\right]^{1/2}, \qquad (8)$$

where $\frac{\partial g_o}{\partial U_{ph}}$; $\frac{\partial g_o}{\partial U_o}$; $\frac{\partial g_o}{\partial U_{o1}}$; $\frac{\partial g_o}{\partial \sin\alpha}$; $\frac{\partial g_o}{\partial \sin\alpha_1}$; $\frac{\partial g_o}{\partial b_o}$; – partial derivatives of the function $g_o = f(U_{ph}, U_o, U_{o1}, \sin\alpha, \sin\alpha_1, b_o)$.

Define the partial derivatives of the function $g_o = f(U_{ph}, U_o, U_{o1}, \sin\alpha, \sin\alpha_1, b_o)$ by variables U_{ph}, U_o, U_{o1}, $\sin\alpha$, $\sin\alpha_1$, b_o:

$$\frac{\partial g_o}{\partial U_{ph}} = -\frac{U_o U_{o1} b_o}{U_{ph}^2(U_o \sin\alpha_1 - U_{o1}\sin\alpha)};$$

$$\frac{\partial g_o}{\partial U_o} = -\frac{U_{o1}^2 b_o \sin\alpha}{U_{ph}(U_o \sin\alpha_1 - U_{o1}\sin\alpha)^2};$$

$$\frac{\partial g_o}{\partial U_{o1}} = \frac{U_o^2 b_o \sin\alpha_1}{U_{ph}(U_o \sin\alpha_1 - U_{o1}\sin\alpha)};$$

$$\frac{\partial g_o}{\partial \sin\alpha} = \frac{U_o U_{o1}^2 b_o}{U_{ph}(U_o \sin\alpha_1 - U_{o1}\sin\alpha)^2};\qquad (9)$$

$$\frac{\partial g_o}{\partial \sin\alpha_1} = -\frac{U_o^2 U_{o1} b_o}{U_{ph}(U_o \sin\alpha_1 - U_{o1}\sin\alpha)^2};$$

$$\frac{\partial g_o}{\partial b_o} = -\frac{U_o U_{o1}}{U_{ph}(U_o \sin\alpha_1 - U_{o1}\sin\alpha)}.$$

Here ΔU_{ph}; ΔU_o; ΔU_{o1}; $\Delta\sin\alpha$; $\Delta\sin\alpha_1$; Δb_o – absolute errors of direct measurements of quantities U_{ph}; U_o; U_{o1}; $\sin\alpha$; $\sin\alpha_1$; b_o, which are defined by the following expressions:

$$\Delta U_{ph} = U_{ph}\Delta U_{ph*};$$

$$\Delta U_o = U_o \Delta U_{o*};$$

$$\Delta U_{o1} = U_{o1}\Delta U_{o1*};\qquad (10)$$

$$\Delta\sin\alpha = \sin\alpha\Delta\sin\alpha_*;$$

$$\Delta\sin\alpha_1 = \sin\alpha_1\Delta\sin\alpha_{1*};$$

$$\Delta b_o = b_o \Delta b_{o*}.$$

To determine the error of measuring devices, we assume that:

$$\Delta U_{ph*} = \Delta U_{o*} = \Delta U_{o1*} = \Delta\sin\alpha_* = \Delta\sin\alpha_{1*} = \Delta b_{o*} = \Delta,$$

where ΔU_{ph*}; ΔU_{o*}; ΔU_{o1*} – the relative error of measuring voltage circuits;
$\Delta\sin\alpha_*$; $\Delta\sin\alpha_{1*}$ – the relative error of the measuring circuits of the angle of lag between the voltage vector of the damaged phase of the electrical network and the line voltage formed by the potentials of the other two phases;
Δb_{o*} – the relative error of the measuring device, which measures the value of additional capacitive conductances that are connected between the phases of the network and the ground.

Solving the equation (8), substituting in it the values of the partial derivatives of the equation (9) and the values of the partial absolute errors (10), assuming that $\Delta U_* = \Delta b_{o*} = \Delta$, we obtain the mathematical dependence of the determination of the random relative mean square error for the method of determining the conductivity, which reduces the insulation resistance between one of the phases of the network and ground in networks with an isolated neutral voltage of more than 1000 V:

$$\varepsilon_{g_o} = \Delta\left[2 + \frac{2(U_{o1}^2\sin^2\alpha + U_o\sin^2\alpha_1)}{(U_o\sin\alpha_1 + U_{o1}\sin\alpha)^2}\right]^{1/2},\qquad (11)$$

or

$$\varepsilon_{g_o} = \Delta\left[2+2\left(1-\frac{2U_oU_{o1}\sin\alpha\sin\alpha_1}{U_{o1}^2\sin^2\alpha+U_{o1}^2\sin^2\alpha_1}\right)^{-1}\right]^{1/2}. \quad (12)$$

We express the resulting equation (12) in relative units and perform its investigation in the Mathcad environment:

$$\varepsilon_{g_o} = \Delta\left[2+\frac{2(\sin^2\alpha_*+U_{o*})}{(U_{o*}+\sin\alpha_*)^2}\right]^{1/2}, \quad (13)$$

where $U_{o*}=\dfrac{U_o}{U_{o1}}$, $\sin\alpha_* = \dfrac{\sin\alpha}{\sin\alpha_1}$.

The relative rms error for capacitive conductivity of insulation in an asymmetric network with an isolated neutral with voltage above 1000 V is determined from the equation (2):

$$b=\frac{U_{o1}b_o\sin\alpha}{U_o\sin\alpha_1 - U_{o1}\sin\alpha},$$

where U_o, U_{o1}, $\sin\alpha$, $\sin\alpha_1$, b_o are the quantities that determine the capacitive conductivity of the insulation in an asymmetric network with an isolated neutral with voltage above 1000 V, obtained by direct measurement.

The relative root-mean-square error in the method for determining the capacitive conductivity of an insulation in an asymmetric network with an isolated neutral with voltage above 1000 V is determined from the expression:

$$\varepsilon_b = \frac{1}{b}\left[\left(\frac{\partial b}{\partial U_o}\right)^2(\Delta U_o)^2+\left(\frac{\partial b}{\partial U_{o1}}\right)^2(\Delta U_{o1})^2+\left(\frac{\partial b}{\partial \sin\alpha}\right)^2(\Delta\sin\alpha)^2 +\left(\frac{\partial b}{\partial \sin\alpha_1}\right)^2(\Delta\sin\alpha_1)^2+\left(\frac{\partial b}{\partial b_o}\right)^2(\Delta b_o)^2\right]^{1/2}, \quad (14)$$

where $\dfrac{\partial b}{\partial U_o}$; $\dfrac{\partial b}{\partial U_{o1}}$; $\dfrac{\partial b}{\partial \sin\alpha}$; $\dfrac{\partial b}{\partial \sin\alpha_1}$; $\dfrac{\partial b}{\partial b_o}$ – are partial derivatives of the $b = f(U_o, U_{o1}, \sin\alpha, \sin\alpha_1, b_o)$.

Define the partial derivatives of the function $b = f(U_o, U_{o1}, \sin\alpha, \sin\alpha_1, b_o)$ by variables $U_o, U_{o1}, \sin\alpha, \sin\alpha_1, b_o$:

$$\frac{\partial b}{\partial U_o} = -\frac{U_{o1}b_o\sin\alpha\sin\alpha_1}{(U_o\sin\alpha_1 - U_{o1}\sin\alpha)^2};$$
$$\frac{\partial b}{\partial U_{o1}} = \frac{U_o b_o\sin\alpha\sin\alpha_1}{(U_o\sin\alpha_1 - U_{o1}\sin\alpha)^2}; \quad (15)$$

$$\frac{\partial b}{\partial \sin\alpha} = \frac{U_o U_{o1} b_o \sin\alpha_1}{(U_o \sin\alpha_1 - U_{o1} \sin\alpha)^2};$$
$$\frac{\partial b}{\partial \sin\alpha_1} = -\frac{U_o U_{o1} b_o \sin\alpha}{(U_o \sin\alpha_1 - U_{o1} \sin\alpha)^2};$$
$$\frac{\partial b}{\partial b_o} = \frac{U_{o1} \sin\alpha}{U_o \sin\alpha_1 - U_{o1} \sin\alpha}.$$

Here ΔU_o; ΔU_{o1}; $\Delta\sin\alpha$; $\Delta\sin\alpha_1$; Δb_o – absolute errors of direct measurements of quantities U_o; U_{o1}; $\sin\alpha$; $\sin\alpha_1$; b_o,which are defined by the following expressions:

$$\begin{aligned} \Delta U_o &= U_o \Delta U_{o*}; \\ \Delta U_{o1} &= U_{o1} \Delta U_{o1*}; \\ \Delta\sin\alpha &= \sin\alpha \Delta\sin\alpha_*; \\ \Delta\sin\alpha_1 &= \sin\alpha_1 \Delta\sin\alpha_{1*}; \\ \Delta b_o &= b_o \Delta b_{o*}. \end{aligned} \quad (16)$$

To determine the error of measuring instruments, we assume that:

$$\Delta U_{o*} = \Delta U_{o1*} = \Delta\sin\alpha_* = \Delta\sin\alpha_{1*} = \Delta b_{o*} = \Delta,$$

where ΔU_{o*}; ΔU_{1o*} – relative error of measuring voltage circuits;
$\Delta\sin\alpha_*$; $\Delta\sin\alpha_{1*}$ – relative error of the measuring circuits of the phase angle between the voltage vector of the damaged phase of the electrical network and the line voltage formed by the potentials of the other two phases;
Δb_{o*} – the relative error of the measuring device, which measures the value of additional capacitive conductances, which are connected between the phases of the network and the ground.

Solving the equation (14), substituting in it the values of the partial derivatives of equation (15) and the values of the partial absolute errors (16), assuming that $\Delta U_* = \Delta b_{o*} = \Delta$, we obtain the mathematical dependence of the determination of the random relative mean square error for the method for determining the capacitive conductivity of insulation in an asymmetric network with an isolated neutral with voltage above 1000 V:

$$\varepsilon_b = \Delta \left[1 + \frac{4U_o^2 \sin^2\alpha_1}{(U_o \sin\alpha_1 + U_{o1} \sin\alpha)^2} \right]^{1/2}. \quad (17)$$

We express the resulting equation (17) in relative units and perform its investigation in the MathCAD environment:

$$\varepsilon_b = \Delta \left[1 + \frac{4U_*^2}{(U_* + \sin\alpha_*)^2} \right]^{1/2}. \quad (18)$$

where $U_{o*} = \dfrac{U_o}{U_{o1}}$, $\sin\alpha_* = \dfrac{\sin\alpha}{\sin\alpha_1}$.

The relative rms error for the total conductivity of the insulation in an asymmetric network with an isolated neutral with voltage above 1000 V is determined from the mathematical dependence (6):

$$y = \frac{U_{pho} U_{o1} b_o}{U_{ph}(U_o \sin\alpha_1 - U_{o1} \sin\alpha)},$$

where U_{ph}, U_o, U_{o1}, $\sin\alpha$, $\sin\alpha_1$, b_o – are the values that determine the total conductivity of the insulation in an asymmetric network with an isolated neutral with voltage above 1000 V, obtained by direct measurement.

The relative root-mean-square error of the method for determining the total conductivity of an insulation in an asymmetric network with an isolated neutral with voltage higher than 1000 V is determined from the expression:

$$\varepsilon_y = \frac{1}{y}\left[\begin{array}{l}\left(\frac{\partial y}{\partial U_{ph}}\right)^2(\Delta U_{ph})^2+\left(\frac{\partial y}{\partial U_{pho}}\right)^2(\Delta U_{pho})^2+\left(\frac{\partial y}{\partial U_o}\right)^2(\Delta U_o)^2+\left(\frac{\partial y}{\partial U_{o1}}\right)^2(\Delta U_{o1})^2\\+\left(\frac{\partial y}{\partial \sin\alpha}\right)^2(\Delta\sin\alpha)^2+\left(\frac{\partial y}{\partial \sin\alpha_1}\right)^2(\Delta\sin\alpha_1)^2+\left(\frac{\partial y}{\partial b_o}\right)^2(\Delta b_o)^2\end{array}\right]^{1/2}, \quad (19)$$

where $\frac{\partial y}{\partial U_{ph}}$; $\frac{\partial y}{\partial U_{pho}}$; $\frac{\partial y}{\partial U_o}$; $\frac{\partial y}{\partial U_{o1}}$; $\frac{\partial y}{\partial \sin\alpha}$; $\frac{\partial y}{\partial \sin\alpha_1}$; $\frac{\partial y}{\partial b_o}$ – partial derivatives of the function $y = f(U_{ph}, U_o, U_{o1}, \sin\alpha, \sin\alpha_1, b_o)$.

Define the partial derivatives of the function $y = f(U_{ph}, U_o, U_{o1}, \sin\alpha, \sin\alpha_1, b_o)$ by variables U_{ph}, U_o, U_{o1}, $\sin\alpha$, $\sin\alpha_1$, b_o:

$$\begin{aligned}
\frac{\partial y}{\partial U_{ph}} &= -\frac{U_{pho}U_{o1}b_o}{U_{ph}^2(U_o\sin\alpha_1 - U_{o1}\sin\alpha)};\\
\frac{\partial y}{\partial U_{pho}} &= -\frac{U_{o1}b_o}{U_{ph}(U_o\sin\alpha_1 - U_{o1}\sin\alpha)};\\
\frac{\partial y}{\partial U_o} &= -\frac{U_{pho}U_{o1}b_o\sin\alpha_1}{U_{ph}(U_o\sin\alpha_1 - U_{o1}\sin\alpha)^2};\\
\frac{\partial y}{\partial U_{o1}} &= \frac{U_{pho}U_o b_o\sin\alpha_1}{U_{ph}(U_o\sin\alpha_1 - U_{o1}\sin\alpha)^2};\\
\frac{\partial y}{\partial \sin\alpha} &= \frac{U_{pho}U_{o1}^2 b_o}{U_{ph}(U_o\sin\alpha_1 - U_{o1}\sin\alpha)^2};\\
\frac{\partial y}{\partial \sin\alpha_1} &= -\frac{U_{pho}U_o U_{o1}b_o}{U_{ph}(U_o\sin\alpha_1 - U_{o1}\sin\alpha)^2};\\
\frac{\partial y}{\partial b_o} &= -\frac{U_{pho}U_{o1}}{U_{ph}(U_o\sin\alpha_1 - U_{o1}\sin\alpha)}.
\end{aligned} \quad (20)$$

Here ΔU_{ph}; ΔU_o; ΔU_{o1}; $\Delta\sin\alpha$; $\Delta\sin\alpha_1$; Δb_o – absolute errors of direct measurements of quantities U_{ph}; U_o; U_{o1}; $\sin\alpha$; $\sin\alpha_1$; b_o, which are defined by the following expressions:

$$\begin{aligned}
&\Delta U_{ph} = U_{ph}\Delta U_{ph*};\ \Delta U_o = U_o\Delta U_{o*};\ \Delta U_{o1} = U_{o1}\Delta U_{o1*};\\
&\Delta\sin\alpha = \sin\alpha\Delta\sin\alpha_*;\ \Delta\sin\alpha_1 = \sin\alpha_1\Delta\sin\alpha_{1*};\ \Delta b_o = b_o\Delta b_{o*}.
\end{aligned} \quad (21)$$

To determine the error of measuring devices, we assume that:

$$\Delta U_{ph*} = \Delta U_{o*} = \Delta U_{o1*} = \Delta \sin\alpha_* = \Delta \sin\alpha_{1*} = \Delta b_{o*} = \Delta,$$

where ΔU_{ph*}; ΔU_{o*}; ΔU_{o1*} – relative error of measuring voltage circuits;

$\Delta \sin\alpha_*$; $\Delta \sin\alpha_{1*}$ – the relative error of the measurement circuits of the phase angle between the voltage vector of the damaged phase of the electrical network and the line voltage formed by the potentials of the other two phases;

Δb_{o*} – the relative error of the measuring device, which measures the value of additional capacitive conductances that are connected between the phases of the network and the ground.

Solving equation (19) by substituting in it the values of the partial derivatives of equation (20) and the values of the partial absolute errors (21), assuming that $\Delta U_* = \Delta b_{o*} = \Delta$, we obtain the mathematical dependence of the determination of the random relative mean square error for the method for determining the total conductivity of insulation in an asymmetric network with isolated neutral voltage above 1000 V:

$$\varepsilon_y = \Delta\left[3 + \frac{3U_o^2 \sin^2\alpha_1 + U_{o1}^2 \sin^2\alpha}{(U_o \sin\alpha_1 + U_{o1}\sin\alpha)^2}\right]^{1/2}. \qquad (22)$$

The resulting equation (22) is expressed in relative units and we perform its investigation in the MathCAD environment:

$$\varepsilon_y = \Delta\left[3 + \frac{3U_{o*}^2 + \sin^2\alpha_*}{(U_{o*} + \sin\alpha_*)^2}\right]^{1/2}, \qquad (23)$$

where $U_{o*} = \dfrac{U_o}{U_{o1}}$, $\sin\alpha_* = \dfrac{\sin\alpha}{\sin\alpha_1}$.

On the basis of the obtained results of random relative mean square errors of determining the active conductivity, which reduces the isolation level between one of the phases of the network and ground, and the capacitive, active and complete conductivities of insulation in a symmetrical network with an isolated neutral voltage above 1000 V, under operating voltage, construct the dependencies:

$$\varepsilon_{g_o} = \frac{\Delta b}{\Delta} = f(U_{o*};\ \sin\alpha_*);$$

$$\varepsilon_b = \frac{\Delta b}{\Delta} = f(U_{o*};\ \sin\alpha_*);$$

$$\varepsilon_y = \frac{\Delta b}{\Delta} = f(U_{o*};\ \sin\alpha_*),$$

presented in figure 1, figure 2 and figure 3.

The obtained mathematical dependences of the relative mean square errors, active conductivity, which reduces the isolation level between one of the phases of the network and the ground ε_{g_o}, capacitive ε_b, full ε_y conductivity of phase isolation of an electrical network with an isolated neutral, presented in a graphical design (figure 1, figure 2 and figure 3) , show that the value of the capacitive additional conductivity b_o, which is introduced between the phases of the electrical network and the ground, affects the variation of the error, depending on the reliability of the sought values.

An error analysis shows that in order to provide the required accuracy for each particular network, the value of the capacitive additional conductivity is selected.

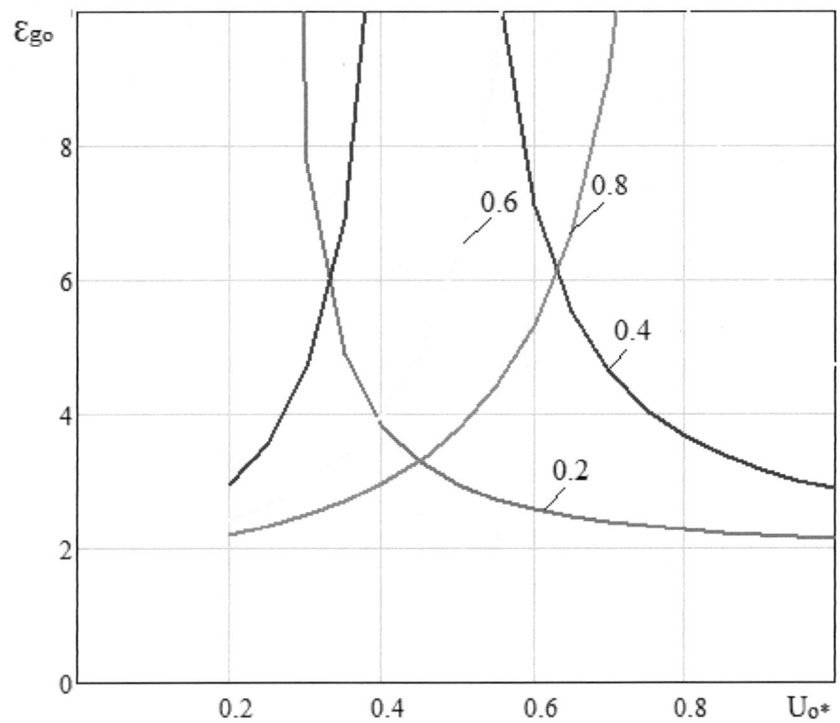

$\sin\alpha_*$=0.2; 0.4; 0.6; 0.8

Figure 1. Relative mean-square errors of determining the active conductivity, reducing the level of insulation between one of the phases of the network and the earth

When determining the active conductivity that reduces the level of isolation between one of the phases of the network and the ground, based on the obtained graphs (figure 1) relative rms error, capacitances forming capacitive conductivity are selected that are connected between phases of the electrical network and ground so that when: $\sin\alpha_*$= 0.2 the zero sequence voltage in relative units was in the range of U_{o*}=0.3-0.9; $\sin\alpha_*$= 0.4, the zero sequence voltage in relative units was in the range of U_{o*}=0.2 to 0.35 or U_{o*}=0.6 to 0.9; $\sin\alpha_*$= 0.6 the zero sequence voltage in relative units was in the range of U_{o*}=0.2-0.5; $\sin\alpha_*$= 0.8 the zero sequence voltage in relative units was within the range of U_{o*}=0.2-0.65.

Then the error does not exceed 10% when using measuring instruments with accuracy of 1.0, and 5% when using measuring instruments with accuracy class of 0.5, and when using measuring instruments with accuracy class of 0.2, an error of not more than 2% is provided.

When determining the capacitive conductivity of the insulation of the phases of the electrical network relative to the ground, based on the error-correction curves shown in figure 2, such capacitances forming capacitive conductivity are selected, which are connected between the phases of the electrical network and the ground, so that at: $\sin\alpha_*$= 0.2 the zero sequence voltage in relative units is within the range of U_{o*}=0.2-0.9; $\sin\alpha_*$= 0.4, the zero sequence voltage in relative units was in the range of U_{o*}=0.2-0.3 or U_{o*}=0.5-0.9; $\sin\alpha_*$= 0.6 the zero sequence voltage in relative units was in the range of U_{o*}=0.2-0.5 or U_{o*}=0.7-0.9; $\sin\alpha_*$= 0.8 the zero sequence voltage in relative

IOP Conf. Series: Materials Science and Engineering **313** (2018) 012015 doi:10.1088/1757-899X/313/1/012015

units was within the range of U_{o*}=0.2-0.65. Then the error does not exceed 10% when using measuring instruments with accuracy class of 1.0, and 5% when using measuring instruments with accuracy class of 0.5, and when using measuring instruments with accuracy class of 0.2 an error of not more than 2% is provided.

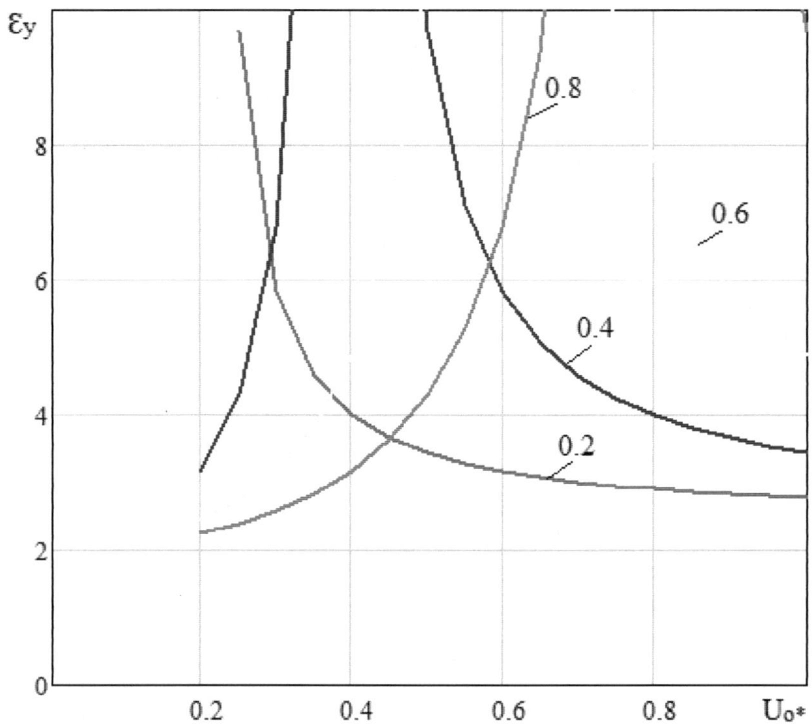

$\sin\alpha_*$=0.2; 0.4; 0.6; 0.8

Figure 2. Relative mean-square errors in determining the capacitive conductivity of a network isolation with an isolated neutral with voltage above 1000 V

When determining the total conductivity of the insulation of the phases of the electrical network relative to the ground, based on the error-correction curves shown in figure 3, capacitances that form capacitive conductivity are selected, which are connected between phases of the electrical network and ground, so that at: $\sin\alpha_*$=0.2 the zero sequence voltage in relative units is within the range of U_{o*}=0.3-0.9; $\sin\alpha_*$= 0.4, the zero sequence voltage in relative units was in the range of U_{o*}=0.2-0.3 or U_{o*}=0.5-0.9; $\sin\alpha_*$= 0.6 the zero sequence voltage in relative units was in the range of U_{o*}=0.2-0.5 or U_{o*}=0.75-0.9; $\sin\alpha_*$= 0.8 the zero sequence voltage in relative units was in the range of U_{o*} =0.2-0.3 or 0.5-0.9. Then the error does not exceed 10% when using measuring instruments with accuracy class of 1.0, and 5% when using measuring instruments with accuracy class of 0.5, and when using measuring instruments with accuracy class of 0.2, an error of not more than 2% is provided.

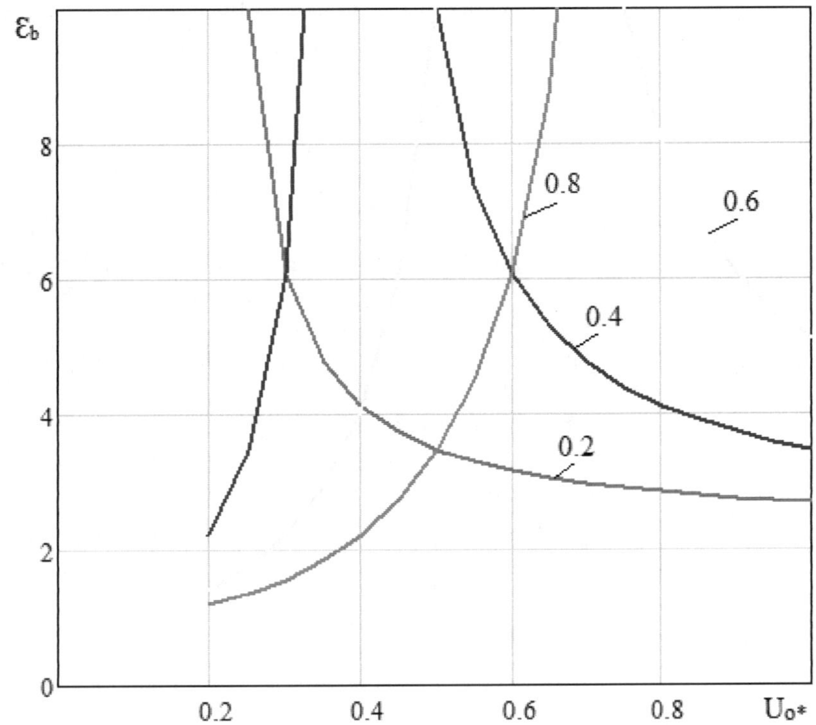

$\sin\alpha_*$=0.2; 0.4; 0.6; 0.8

Figure 3. Relative root-mean-square errors in determining the total conductivity of a network isolation with an isolated neutral with voltage above 1000 V

Since the random relative mean square error of the method for determining the active conductivity of insulation in an asymmetric network with a voltage higher than 1000 V is determined by the Pythagoras formula and can be represented as:

$$\varepsilon_g = \frac{g_*}{\Delta} = \frac{1}{g}\left[\left(\frac{\partial y}{\partial g}\right)^2 \Delta y^2 + \left(\frac{\partial b}{\partial g}\right)^2 \Delta b^2\right]^{1/2}, \qquad (24)$$

then it follows that the error in the correct selection of capacitive additional conductivity that are connected between the phases of the electrical network and the ground in determining for capacitive and complete conductivities of network isolation will be legitimate and for the active conductivity of network isolation.

Therefore, the error in the active conductivity of the network isolation will represent both the geometric sum of the capacitive and total conductivities of the network isolation. On the basis of the foregoing it follows that the error in determining the active conductivity does not exceed 10% when using measuring instruments with an accuracy class of 0.5.

The conducted studies of the random relative mean square errors show that the accuracy of indirect measurements in the developed method can be effectively regulated not only by selection of capacitive additional conductivity, which are connected between phases of the electric network and ground, but also by the selection of measuring devices according to the accuracy class. When choosing meters with an accuracy class of 0.5 with the correct selection of capacitive additional conductivity that are connected between the phases of the electrical network and the ground, the errors in measuring the active conductivity of the insulation of the network will not exceed 10%.

IOP Conf. Series: Materials Science and Engineering **313** (2018) 012015 doi:10.1088/1757-899X/313/1/012015

On the basis of the foregoing, it follows that the method developed provides satisfactory accuracy in determining the insulation parameters in a three-phase unbalanced electrical network with an isolated neutral, as well as the simplicity and safety of operating in existing electrical installations with voltages above 1000 V in industrial and mining enterprises.

3. Conclusion

When introducing a method for determining the active conductivity that reduces the level of isolation between one of the phases of the network and the ground and the insulation parameters in an asymmetric network with a voltage above 1000 V based on the analysis of the random relative mean square error, it is found that in order to ensure the required accuracy for each particular network should be selected its own value of additional conductivity that is included between the phases of the electrical network and the earth, so as to provide the required satisfactory accuracy at definition for:

- active conductivity reducing the level of isolation between one of the phases of the network and the ground, based on the obtained graphs of the relative rms errors, capacitances forming capacitive conductivity are selected that are connected between phases of the electrical network and ground so that at: $\sin\alpha_* = 0.2$, the zero sequence voltage in relative units was within the range of U_{o*}=0.3-0.95; $\sin\alpha_* = 0.4$, the zero sequence voltage in relative units was in the range of U_{o*}=0.2 to 0.35 or U_{o*}=0.6 to 0.95; $\sin\alpha_* = 0.6$ the zero sequence voltage in relative units was in the range of U_{o*}=0.2-0.5; $\sin\alpha_* = 0.8$ the zero sequence voltage in relative units was in the range of U_{o*}=0.2 to 0.65, then the error does not exceed 10% when using measuring instruments with accuracy class of 1.0, and 5% when using measuring instruments with accuracy class of 0.5, and when using measuring instruments with accuracy class of 0.2, an error of not more than 2% is provided;
- capacitive conductivity of the insulation of the phases of the electrical network with respect to the ground, such capacitors forming capacitive conductivity are selected which are connected between the phases of the electrical network and the ground so that at: $\sin\alpha_* = 0.2$ the zero sequence voltage in relative units is within the range of U_{o*}=0.2-0.9 ; $\sin\alpha_* = 0.4$, the zero sequence voltage in relative units was in the range of U_{o*}=0.2-0.3 or U_{o*}=0.5-0.9; $\sin\alpha_* =$ 0.6 the zero sequence voltage in relative units was in the range of U_{o*}=0.2-0.5 or U_{o*}=0.7-0.9; $\sin\alpha_* = 0.8$ the zero sequence voltage in relative units was within the range of U_{o*}=0.2-0.65. Then the error does not exceed 10% when using measuring instruments with accuracy class of 1.0, and 5% when using measuring instruments with accuracy class of 0.5, and when using measuring instruments with accuracy class of 0.2, an error of not more than 2% is provided;
- the total conductivity of the insulation of the phases of the electrical network relative to the ground, such capacitances forming capacitive conductivity are selected that are connected between the phases of the electrical network and the ground so that at: $\sin\alpha_* = 0.2$ the zero sequence voltage in relative units is within the range U_{o*}=0.3-0.9 ; $\sin\alpha_* = 0.4$, the zero sequence voltage in relative units was in the range of U_{o*}=0.2-0.3 or U_{o*}=0.5-0.9; $\sin\alpha_* =$ 0.6 the zero sequence voltage in relative units was in the range of U_{o*}=0.2-0.5 or U_{o*}=0.75-0.9; $\sin\alpha_* = 0.8$ the zero sequence voltage in relative units was in the range of U_{o*}=0.2-0.3 or U_{o*}=0.5-0.9. Then the error does not exceed 10% when using measuring instruments with accuracy class of 1.0, and 5% when using measuring instruments with accuracy of 0.5, and

when using measuring instruments with accuracy class of 0.2, an error of not more than 2% is provided;

- the active conductivity of network isolation the error will represent both the geometric sum of the capacitive and total conductivities of the network isolation. On the basis of the foregoing it follows that the error in determining the active conductivity does not exceed 10% when using measuring instruments with an accuracy class of 0.5.

On the basis of the foregoing, it follows that the method developed provides satisfactory accuracy in determining the insulation parameters in an asymmetric network with an isolated neutral with voltages above 1000 V in industrial and mining enterprises.

References

[1] Gladilin L V, Shchutsky V I, Bacezhev Yu G and Chebotayev N I 1977 *Electrical safety in the mining industry* (Moscow: Nedra) p 327

[2] Shchutsky V I, Mavritsyn A M, Sidorov A I and Sitchichin Yu V 1983 *Electrical safety in open mining operations* (Moscow: Nedra) p 192

[3] Utegulov B B and Abdykarimov A B 1993 The method of determining the value of conductivity, reducing the level of insulation in a network with an isolated neutral with voltage above 1000 V of mining enterprises *Integrated use of mineral raw materials (article on russian language)* **3** 30–34

[4] Utegulov B B 2016 *The method of determining the insulation parameters in networks with a voltage of 6 - 10 kV, V Int. Scientific and Practical Conf. Proc. on Efficient and high-quality supply and use of electricity (article and conf. on russian language)* ISBN: 978-5-8295-0439-7 (Ekaterinburg: publishing house of Ural Federal University named after the first President of Russia B.N. Yeltsin)

[5] Utegulov B B, Utegulov A B and Uakhitova A B 2016 Increasing the efficiency of protective devices for mining machines *Journal of Mining Science (article on russian language)* **2** 102–108

[6] Utegulov B B and Abdykarimov A B 1993 Determination of insulation parameters in the 6 - 10 kV network with a weakened insulation between one of the phases of the network and the ground of mining enterprises *Integrated use of mineral raw materials (article on russian language)* **1** 16–21

[7] Gladilin L V and Utegulov B B 1980 Analysis of the error of the method for determining the insulation parameters in three-phase electrical networks with an isolated neutral with voltage above 1000 V *Mining magazine (article on russian language)* **8** 94–97

[8] Zaidel A N 1968 *Elementary estimates of measurement errors, 3rd ed.* (Leningrad: Science) p 97

[9] Utegulov B B 2016 *The method of determining the insulation parameters in three-phase electrical networks with isolated neutral with voltages up to and above 1000 V (Int. Conf. on New Energy and Future Energy System)* vol 40 (Beijing: IOP Conference Series, Earth and Environmental Science)

[10] Utegulov B B, Utegulov A B and Zhankuanyshev M 2015 Increasing the reliability of power supply and the safety at operation of electroreceivers in the networks of 6-10 kV *Journal of scientific research and development* **2** special issue 7 (September)

[11] Utegulov B B, Utegulov A B and Uakhitova A B 2016 Development of Method to Improve Efficiency of Residual Current Device under 1000 V on Excavators of Mining Enterprises *Journal of Mining Science* **52** 325–331

IOP Conf. Series: Materials Science and Engineering **313** (2018) 012016 doi:10.1088/1757-899X/313/1/012016

Determination of the parameters of the scheme of substitution of the anchor winding of the inductive sup-exciter of the combined multifunctional brushless exciter

N Pavlov[1], V Denisenko[1]

[1] Ural Federal University named after the first President of Russia BN. Yeltsin, Mira Str. 19, RU 620000, Ekaterinburg, Russian Federation
pavlov.nikita@mail.ru

Abstract. The picture of the magnetic field of scattering and mutual induction of the anchor winding of the exciter for the three types of configuration of the tooth-groove layer is considered. The inductive resistance is calculated for the entire phase.

1. Introduction.

With the development of small-scale power generation, there is a growing need to develop new designs of diesel-generator sets that can work in various emergency situations and in extreme climatic conditions, including in fire-hazardous conditions when repairing gas and oil pipelines that require a high degree of reliability of power supply. For operation under such conditions, it is advisable to have a closed version of the generator with a combined brushless exciter (CBE).

One of the pathogens developed at the Department of Electrical Machines of the UrFU is an exciter for a diesel generator (DG) with a capacity of 60 kW, 1500 rpm and located inside the generator housing. In this pathogen, one of the determining functional elements is a combined inductor sub-exciter (ISE). To simplify the manufacturing process in the design of CMBE with a large number of poles, it is advisable to apply a 4-phase anchor winding of the exciter (AWE) [1], which is located in the tooth layer of the polepiece of the synchronous exciter. Technological design of the AWE structure and ISE power, developed both in nominal and forcing mode, largely depends on the structure of the tooth-groove construction and the layout of the pole coil groups of the nuclear reactor. The choice of structure can be made on the basis of a comparative evaluation of the indicated capacities under the same operating conditions of the ISE. As is known, the maximum value of the ISE power is determined by the equivalent inductive resistance of the substitution circuit of the exciter, an essential part of which is the inductive reactance of the scattering of AWE [2].

In this regard, in this article, an estimation of inductive scattering of AWE is carried out with three different variants of the construction of the notch-groove layer.

In figure 1 shows the variants for the construction of the notch-groove layer and the variants for placing the pole coil groups of adjacent phases of the ISE winding, which are used in industrial CBE samples for medium- and high-power generators (variant 1b) and in the bigarm exciton pilot for powerful synchronous motors figure 1, c. In DG pathogens under conditions of restriction on the outer diameter of the inductor, it is advisable to apply a 4-phase double-row winding (figure 1, a) [1]. This winding is placed in the notch-groove layer, which is formed by removing from the circuit of figure 1, c of the outer teeth and teeth having a small width.

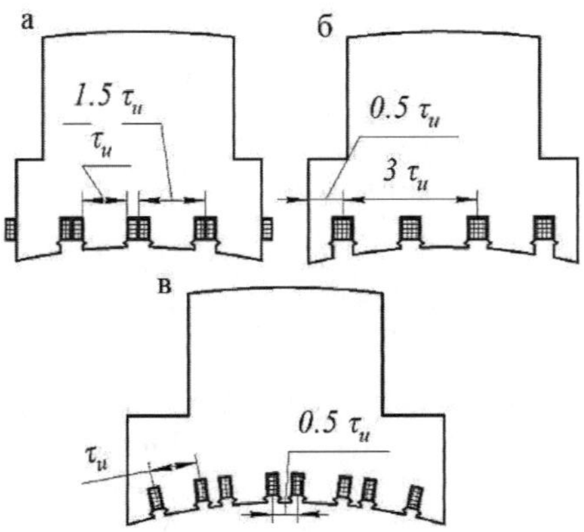

Figure 1. - Various variants of the construction of the tooth-groove layer of the pole piece and methods for constructing the ISE winding.

2. Formulation of the problem

The purpose of the experiment is to find the inductive phase resistance for three different configurations of the tooth-groove layer. The investigation of the magnetic field was carried out by the finite element method.

The current density for all three variants is **j** = 220 A/m2. The number of turns in the phase **w** = 12. The number of turns in one groove for the variant of figure 1, b is 2 times more than for other options. The winding is concentrated **q** = 1. The frequency of the inductor EMF is **f** = 1000 Hz. The air gap is $\delta = 0$. The length of the machine is **l** = 50 mm. The magnetic flux on the back of the inductor is equal to 0 V/m.

In order to simplify the task, the coils of each phase are modeled in such a way that one equivalent loop is located in the groove instead of the calculated number of turns. The cross-section of the equivalent loop is such that the coil creates the same MDS as for a real number of turns.

The calculation of the coefficient of the channel conductivity of the scattering was carried out for the circuits in figure 1, a and figure 1, b by a numerical method and analytic for the circuit in figure 1, b, separately for each section of the groove of each configuration of the tooth-groove layer. We break each slot into 2 sections: 1) the space occupied by the coil of the investigated phase with a given current density 2) the rest of the slot, the field lines of force in which do not pass through the regions occupied by the coils of the phase under investigation.

To determine the conductivity by the numerical method for the circuits in figure 1, a and figure 1, b, a calculation of the magnetic field of the slot scattering of the NRWR was performed. The problem is solved by a pair of poles. Since a mutually inductive coupling between adjacent phases over the scattering field is practically absent in the 4-phase SNR [5], the calculation of the field is made when feeding only the investigated phase.

In accordance with the pattern of the magnetic field, the groove was divided into tubes of equal magnetic flux (regions bounded by two adjacent lines of force). Based on the formula for calculating the coefficients of conductivity of slotted scattering:

IOP Conf. Series: Materials Science and Engineering **313** (2018) 012016 doi:10.1088/1757-899X/313/1/012016

$$\lambda_S = \int_0^{h_S} (\frac{S_X}{S_S})^2 \frac{dx}{b_x} \tag{1}$$

where S_s and S_x are the cross-sectional areas of the groove occupied by the winding conductors; b_x, l_x - width and length of the power line, respectively.

The conductivity calculation method is similar to the method described in [3].

By analytical method, the coefficient of slot conductivity of scattering was determined as the average of the sum of the slots filled with a coil of one phase; The calculation was carried out in the middle of the groove for the circuit in Figure. 1b. by the standard formula [4]:

$$\lambda_S = \lambda_{S1} + \lambda_{S2} \tag{2}$$

$$\lambda_{S1} = \frac{h_S}{3b_S} \tag{3}$$

$$\lambda_{S2} = \frac{h_K}{b_S} \tag{4}$$

where λ_{S1} - the conductivity of the first section; λ_{S2} - groove conductivity of the second section; h_S – the depth of space occupied by the coil; h_k - the thickness of the wedge, the height of the slot, the insulation under the wedge; b_s - is the width of the groove.

In the case of the circuit in figure 1, a, an analytical method for determining the coefficient of slot conductivity of scattering was not used, since there are no analytical expressions for the side of the coil in the interpolar space.

The coefficient of magnetic conductivity of the frontal scattering and the coefficient of magnetic conductivity of the scattering along the crowns of the teeth were calculated by the standard formulas [4].

The calculation of the inductive scattering resistance was carried out according to the standard formula [4]:

$$x_\sigma = 15,8 \frac{f}{100} (\frac{w}{100})^2 \frac{l}{pq} (\lambda_S + \lambda_l + \lambda_k) \tag{5}$$

The calculation of the inductive resistance of mutual induction was also carried out by analytical and numerical methods.

Analytic method, the resistance of mutual induction was calculated by the formula:

$$x_m = 2\pi f w^2 \lambda_m \tag{6}$$

$$\lambda_m = \mu_0 \frac{y_k}{8\delta``} \tag{7}$$

where λ_m is the coefficient of magnetic conductivity of mutual induction; y_k is the step of the coil; $\delta``$ -

equivalent air gap.

By numerical method, the resistance of mutual induction was calculated by formula (6):

$$\lambda_m = \frac{\Phi_m}{F} \tag{8}$$

where Φm is the flux of mutual induction;
F is the magnetizing force.

Inductive resistance of the winding phase was calculated by the formula:

$$x_{ph} = x_\sigma + x_m \tag{9}$$

3. Results and Conclusions

The pattern of the magnetic scattering field is shown in figures 2, 3 and 4, the magnetic field of mutual induction - in figures 5, 6 and 7 for the circuits in figure 1, a, figure 1, b and figure 1, c respectively.

The results of calculating the coefficient of shear conductivity of scattering are presented in Table 1. The results of calculating the inductive resistances are presented in Table 2.

Table 1. Results of calculation of the coefficient of slot conductivity of scattering.

Scheme	λ_{S1}		λ_{S2}	
	Analyte. method	Numbers method	Analyte. method	Numbers method
Figure 1, a	-	0,425	-	0,459
Figure 1, b	0,407	0,507	0,56	0,468
Figure 1, c	0,674	0,659	0,577	0,597

Based on the calculations in Table. 1, the use of standard formulas for calculating the coefficient of shear conductivity of scattering introduces a large error. In this regard, it is necessary to introduce a correction factor (see Table 3) for each section of the groove.

It is advisable to write the expression for the calculation of the conductivity of slotted scattering in the following form:

$$\lambda_S = k_{c1}\frac{h_S}{3b_S} + k_{c2}\frac{h_k}{b_S} \tag{10}$$

where k_{c1} is the correction factor for section 1; k_{c2} is the correction factor for section 2.

As studies have shown, the application of the standard formula for calculating the slot conductivity of scattering for the circuit in Figure. 1b, which was used in the biharmonic exciter, the use of the standard formula does not introduce a significant error and it can be used to calculate the coefficient of

IOP Conf. Series: Materials Science and Engineering **313** (2018) 012016 doi:10.1088/1757-899X/313/1/012016

magnetic conductivity of slotted scattering without introducing correction coefficients. The calculation of the conductivity of the slot scattering for the circuit in Figure. 1b can be performed only with the introduction of correction factors in the standard formula. For the circuit in Figure. 1a it is necessary to obtain an analytical expression for the coefficient of slot conductivity of scattering for the side of the coil located in the interpolar space.

The calculation of the inductive resistance of mutual induction showed that the difference between the analytical method and the numerical method is insignificant. Also, the calculation showed that the inductive resistance of mutual induction is smaller than the inductive scattering resistance. The mutual resistance of the circuits in Figure. La and Figure. 1c to the resistance for the circuit in Figure. 1b, determined analytically, do not differ much from analogous relations determined by numerical methods (see Table 4).

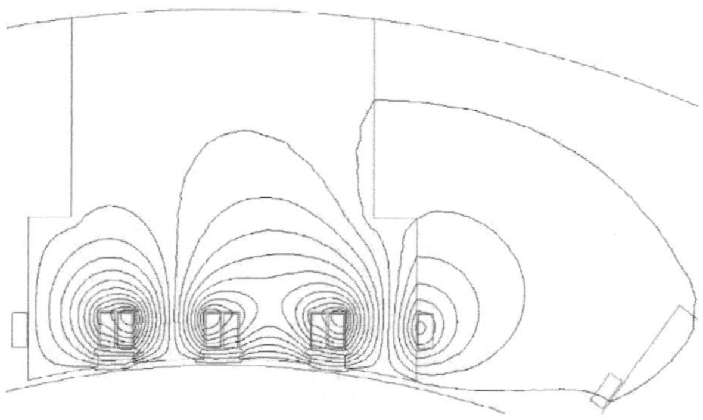

Figure 2. Scattering magnetic field pattern.

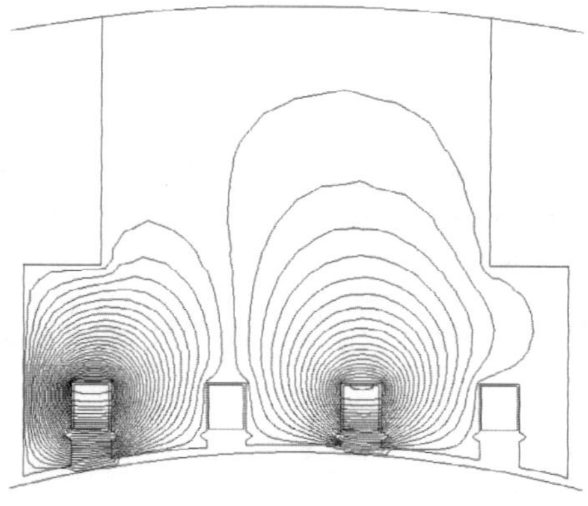

Figure 3. Scattering magnetic field pattern.

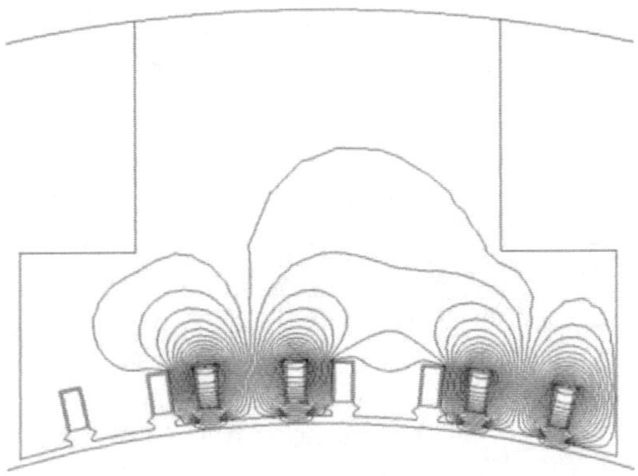

Figure 4. Scattering magnetic field pattern.

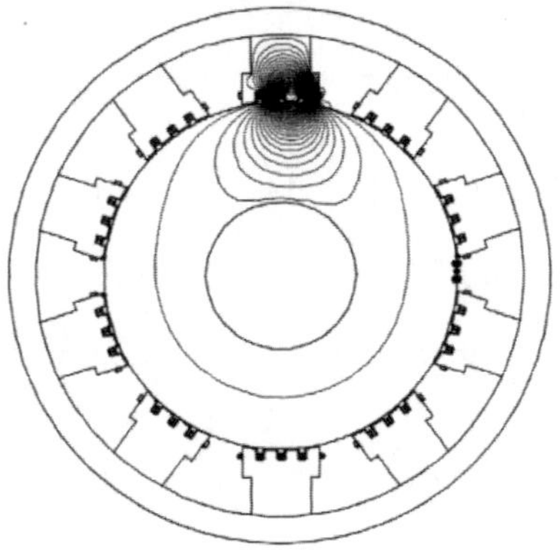

Figure 5. A picture of the magnetic field of mutual induction.

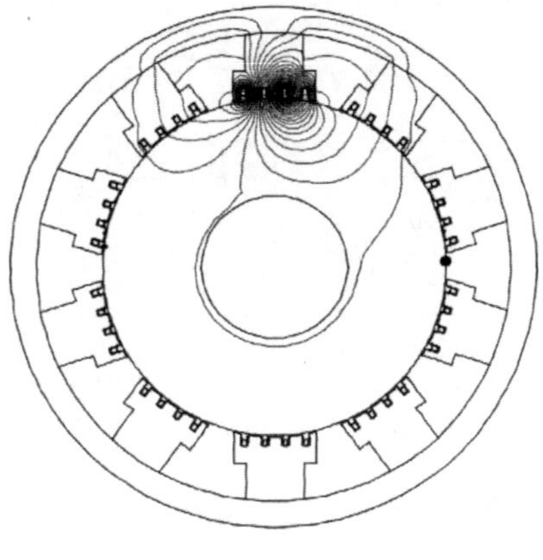

Figure 6. A picture of the magnetic field of mutual induction.

IOP Conf. Series: Materials Science and Engineering **313** (2018) 012016 doi:10.1088/1757-899X/313/1/012016

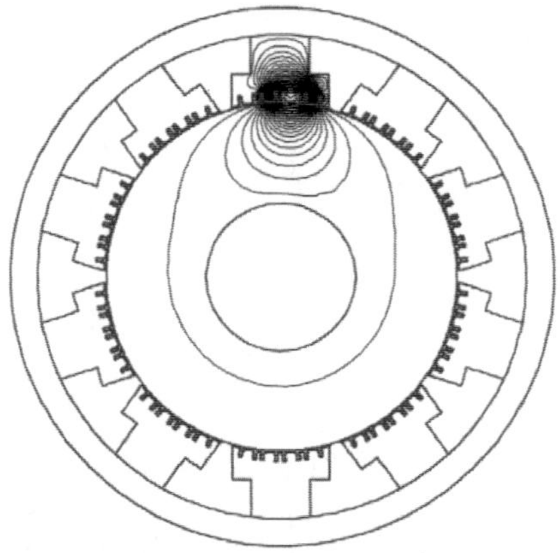

Figure 7. A picture of the magnetic field of mutual induction.

Table 2. Results of calculation of inductive resistances.

		Figure 1, a	Figure 1, b	Figure 1, c
x_σ	Analyte method	-	0,039	0,040
	Numbers method	0,039	0,0399	0,041
x_m	Analyte method	0,011	0,022	0,0079
	Numbers method	0,011	0,024	0,0074
x_{ph}	Analyte method	-	0,061	0,0479
	Numbers method	0,05	0,0639	0,0484

Table 3. Correction factor value.

Symbol	Site	Value
k_{c1}	Site 1	1,246
k_{c2}	Site 2	0,836

Table 4. Ratio of inductive resistances of mutual induction.

Method	x_{m1a}/x_{m1b}	x_{m1c}/x_{m1b}
Analytical method	0,5	0,36
Numbers method	0,46	0,31

REFERENCES

[1] Kychanov M. V. Design features and prospects of closed-circuit diesel-generator sets using nanostructured insulating materials / M. V. Kychanov, A. N. Moiseichenkov, V. I. Denisenko, A. T. Plastun // Scientific Works of the XVI Ural International Conference of Young Scientists on Priority Areas for the Development of Science and Technology: a Collection of Articles. In 3 hours Ekaterinburg: USTU - UPI, 2009. P.1. Pp. 225-228.

[2] Kychanov M. V. Evaluation of the power of the exciter of a combined multifunctional brushless exciter / M.V. Kychanov, A. N. Moysechenkov, V. I. Denisenko, A. T. Plastun // Effective and qualitative supply and use of electricity: a collection of reports 4th International Conference within the framework of the exhibition "Energy Saving, Heating, Ventilation, Water Supply" (Ekaterinburg, May 26-28, 2015). - Ekaterinburg: UrFU, 2015. C. 175-178. ISBN 978 5 321 002284 9.

[3] Kychanov M. V. To calculation of the conductivity of the slotted double-row winding of a combined multifunctional brushless exciter / M. V. Kychanov, P. L. Myalitsin, A. A. Leonov, V. I. Denisenko, // Scientific works of the XVI Ural an international conference of young scientists on priority areas for the development of science and technology: a collection of articles. In 3 hours Ekaterinburg: USTU - UPI, 2009. P.1. Pp. 212-215.

[4] I. P. Kopylov. Designing electrical machines: a textbook for universities / I. P. Kopylov, B. K. Klokov, V. P. Morozkin, B. F. Tokarev; E. d.. I. P. Kopylova.-3rd ed., Rev. and additional. – M .: Higher School, 2002. -757 p.

[5] V. I. Denisenko, "Evaluation of the influence of phase mutual induction in calculating the scattering parameters of a two-row winding of an inductor exciter" / V. I. Denisenko, M. V. Kychanov, P. L. Myalitsin // Problems and achievements in industrial energy: Proceedings VII scientific and practical conference with international participation in the exhibition "Energy and Electrical Engineering. Automated systems and instrument making. Lighting Engineering », November 11 - 13, 2008 Ekaterinburg: Ural Exhibitions - 2000, 2008. - P. 134-138

IOP Conf. Series: Materials Science and Engineering **313** (2018) 012017 doi:10.1088/1757-899X/313/1/012017

The development and testing of a linear induction motor being fed from the source with limited electric power

V V Tiunov

Professor of Electrical Engineering and Electro-mechanics Department, Perm National Research Polytechnic University, 29 Komsomolskiy Prospect, Perm, 614990, Russia

tiuvas@mail.ru

Abstract. The report provides results of the research related to the tubular linear induction motors' application. The motors' design features, a calculation model, a description of test specimens for mining and electric power industry are introduced. The most attention is given to the single-phase motors for high voltage switches drives with the usage of inexpensive standard single-phase transformers for motors' power supply. The method of the motor's parameters determination, when the motor is being fed from the transformer, working in the overload mode, was described, and the results of it practical usage were good enough for the engineering practice.

1. Introduction

Linear Induction Motors (LIMs) directly convert electric power into forward motion movement of working machines or its separate parts [1-3].

Electric Drives with LIMs in many cases allow effectively automatize and mechanize of different technological processes, especially in conditions of aggressive and toxic medium.

Direct conversion of the electromagnetic field power into forward motion movement of secondary bodies gives LIMs great advantages in comparison with equipment which use mechanic, hydraulic or pneumatic (air powered) converters, and also with solenoids and electromagnets, which usually have a limited movement.

The lack of gaseous or liquid subsidiary working body is especially inconvenient under conditions of toxic or radioactive surrounding environment, as in this case there is a necessity to use extra purification systems, and so it is necessary to have special hermetic reservoirs, as well, for keeping used toxic liquid or radioactive air.

Taking into account the construction of LIMs, they can be subdivided into two main types – flat and cylindrical (tubular) ones. Nowadays, the flat LIMs are more investigated and they have rather widespread application in transport and technological systems, and in magneto hydrodynamic pumps. The tubular LIMs (TLIMs) are much less investigated theoretically, because the correct calculations of them are connected with using of special mathematical functions and, therefore, they are more complicated objects for theoretical designing and investigations, than the flat LIMs.

Moreover, the usage of TLIMs can be highly effective solution of a number of problems while working out of different acting mechanisms, pushers, systems of carrying pipes, cylindrical containers and so on, especially for application in aggressive conditions. At the same time, electric drive with TLIMs is rather simple, compact and safe and, above all, inductors and working bodies of the motors can be easily put into hermetic casing, what is extremely important when LIMs are working in conditions of chemical aggressive medium or outdoor.

In this report some examples of TLIMs for mining industry and power electro energetics are shown. Some of the motors developed have been successfully tested by the customers and introduced into production manufacture. The great attention was paid to the most urgent task of modern power industry, which was connected with the further increasing of the power sources reliability.

IOP Conf. Series: Materials Science and Engineering **313** (2018) 012017 doi:10.1088/1757-899X/313/1/012017

One of the key aspects for the technical implementation of this task is the automated distribution of electrical power through the usage of the partitioned grids with back-up lines, which are equipped with the sufficiently simple, reliable and inexpensive shut-off and diverter switching hardware.

Using TLIMs, we can significantly simplify the layout of the switchgear, reduce its dimensions and weight, reduce material consumption and increase its performances and reliability. Besides, such a switching device equipped with the power supply (step-down transformer) can be placed high off the ground at the level of the power transmission supports, hence, increasing external damage resistance. It is important, that switching and other devices, based on the TLIMs, are simple in structure and they can work in three-phase and in single-phase networks of low or high (through the inexpensive transformer) voltage.

2. Design features

Electric machines with linear or reciprocating movement of the working body also became widespread application in the mining and processing industry, in particular, in potash ore mining enterprises and mineral fertilizers producers. These are various dampers, valves, tappets, emergency locks and tech clips, shut-off valves, flow dividers, samplers, etc. [1- 6, 8].

The linear induction motors, designed for some of these applications at the Electrical Engineering and Electro-mechanics Department of the Perm National Research Polytechnic University, are shown in Figures 1, 2.

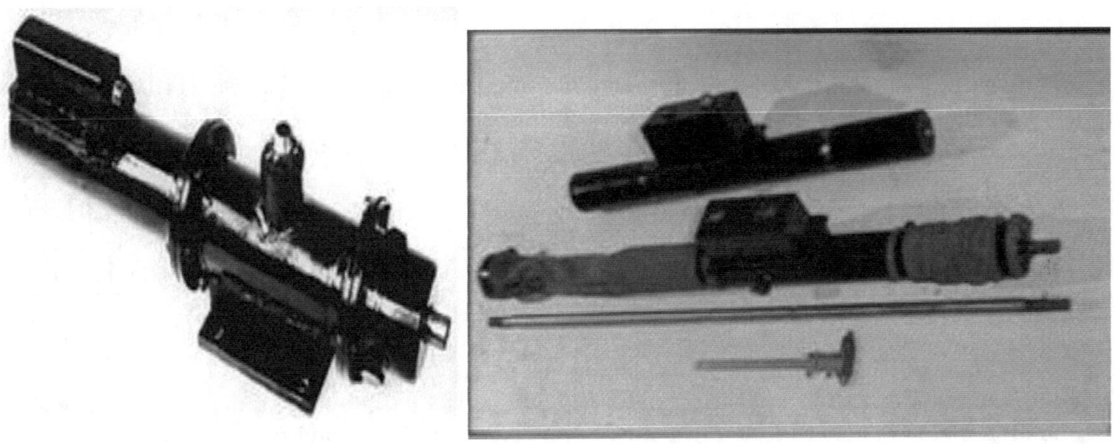

Figure 1. The tubular linear induction drives for mining industry.

Currently, actuation of such mechanisms is effected by a geared motor equipped with a mechanical converter of rotational-to-linear motion of the working body at the outlet of the converter. Hands-on experience demonstrates that these Linear Actuators (LAs), for example of PTV-type (in Russian- 'ПТВ', it means - the electro-mechanical drive of explosion-roof execution), are not sufficiently reliable and are not easy to maintain (especially when operating in aggressive saline environment), and, at the same time, they have a very significant size, weight and cost. In addition, in some cases, response performance (speed of acting) of these linear actuators is unsatisfactory.

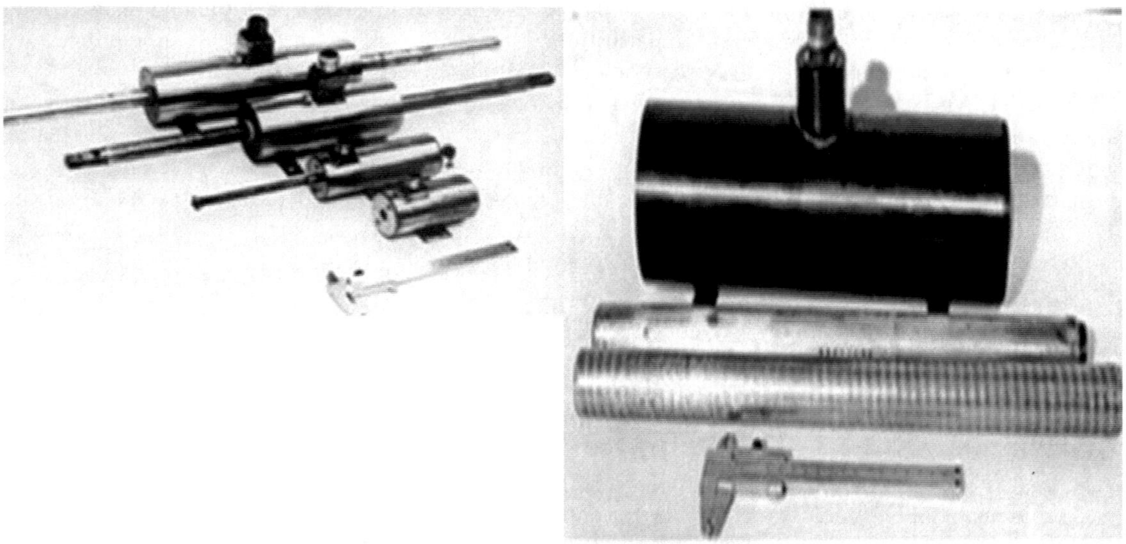

Figure 2. The tubular linear induction motors for switching equipment and for other applications.

3. Description of the TLIMs' design

Aiming to find a solution for this issue of creating a linear actuator on the basis of TLIMs for power and mining industry, we have designed the series of three-phase and single-phase TLIMs of 50-1000 N pulling force (see, for example, Figures 1, 2). Its design features and characteristics were presented in the report.

It is known, that projecting of TLIMs for such linear actuators is rather complicated, as it is conditioned by the influence of edge effects, appearing in the LIM, and also by the great variety of the motors' constructive executions. To take into account the above mentioned peculiarities of TLIMs, the calculation method of its electromagnetic field and integral characteristics was developed based on the usage of presentation about anisotropic media. This calculation method allows getting comparatively simple solutions available for engineering calculations.

The distribution of electromagnetic field in the TLIMs because of the complex geometric configuration of the motor, and the presence of slots on the inductor and on the secondary body as well, is extremely difficult. Because of irregularity of electric and magnetic peculiarities of the TLIMs in axis, the saturation of the motor's ferromagnetic materials influences greatly on the character of magnetic field. If magnetic fields of the flat LIMs are described in aspects of exponential functions [3], then the fields of the TLIMs are more complex and are defined more often by means of sums of special cylindrical functions of Bessel.

To get enough simple calculation correlations, describe the field of the TLIM, it is necessary to define the exact ideal task which allows considerably simplify mathematic correlations without distorting the main physical processes in the motor. In this case, the real motor is usually substituted by the simplified electromagnetic calculating model. While investigating the distribution of electromagnetic field of the TLIM, it is expedient to use calculation model which was offered and described in [4] and it is introduced in Figure 3. This calculation model was used for design of some of the motors described above.

This report specifically provides the results of the single-phase TLIMs elaboration for the high voltage switches drives with usage for its power supply inexpensive single-phase transformers, for example, of the 'OM' - type transformers (in Russian: 'OM'-type means: single-phase-'O', oil-cooled-

IOP Conf. Series: Materials Science and Engineering **313** (2018) 012017 doi:10.1088/1757-899X/313/1/012017

'M', power transformer with natural circulation of oil and air) and, for example, of the 'NOL'-type transformers (in Russian: 'НОЛ' mean the inductive voltage power transformers of the 'NOL'-type, single phase, cast resin, size 08-10.05.1) or their equivalents, being fed from the networks of 6.3 or 10 kV in the short-term overload mode (at the time of tripping power lines). This allows us to significantly simplify the construction and reduce the cost of complete package of linear actuators for different purposes and, in particular, for switching devices.

The common single-phase oil transformers of the 'OM'- series (with a power capacity of 1.25-4.0 kVA, with the High Voltage/Low Voltage -HV/LV-of 10; 6.3/0.23 kV, frequency of 50 Hz), are mainly intended to supply power to the rail-road locking equipment and electrical power centralized control. A scheme and windings' connection group of the transformers is a star and 1/1-0. They have a different climate design for outdoor and for indoor installation with the ambient temperatures from minus 50°C to plus 45°C. The transformers are manufactured as leak-tight items due to valves in a body design, and they have a special bracket for installation at the power line supports.

The inductive voltage transformers 'NOL 08-10.05.1'- type (single phased, cast resin) are designed to supply power to instruments and vacuum circuit breaker actuators of the 'HVTP'- type ('High-Voltage Three-Pole' actuator), and also for installation in mining electrical switchgears with the voltage of 10 kV for the underground potash mines networks. Theirs data: voltages of the HV/LV windings are of 10/0.1; 0.127 kV; the power capacity is of 630 VA or more; the frequency is of 50 Hz; the scheme and winding connection group is the star of 1/1-0. The windings and magnetic circuit are encapsulated in the epoxy resin to form a robust monolithic block which includes threaded sleeves for the transformer erection at the installation site.

Our calculation model of a tubular linear induction motor is shown in Figure 3. This model was being used for the TLIMs' design.

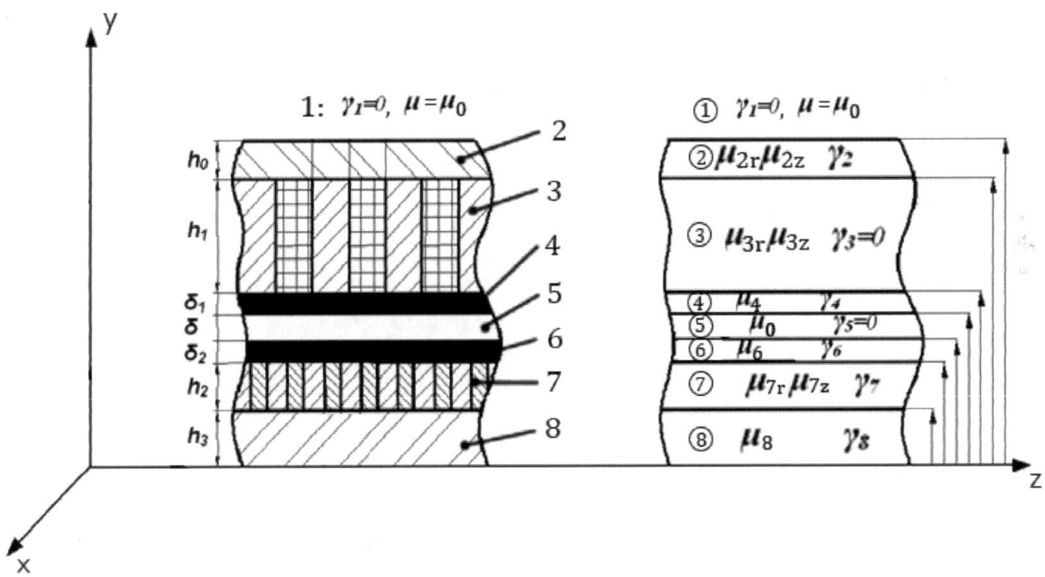

Figure 3. The calculation model of a tubular linear induction motor:

1- a zone of the surrounding air;
2- a zone of the frame and inductor's core;
3- a zone of the slots and teeth structure of an inductor;
4- a zone of the inner case of an inductor;
5- a zone of the air gap between inductor and working body;
6- a zone of the working body's cover;
7- a zone of the slots and teeth structure of a working body;
8- a zone of an inner working body's core.

IOP Conf. Series: Materials Science and Engineering **313** (2018) 012017 doi:10.1088/1757-899X/313/1/012017

To verify the single-phase capacitor's TLIMs efficiency to be used for high voltage power disconnector, TLIM was supplied from the limited power source, namely OM-type transformer (single-phase oil transformer with natural circulation of oil and air). We used one of the tubular single-phase linear motor with length and diameter of respectively 256 mm and 72 mm. The motors had a variety of combined working bodies: copper-plated secondary ferromagnetic element; secondary ferromagnetic element with the 'twisted' (rolled up or ring-shaped) squirrel cage in a ferromagnetic array, etc. All elements had the outer diameter of 20 mm. During the process of an experiment, in the laboratory conditions we measured pulling force (F), currents in the both phases of the two-phase winding ($I_{A.\ B.}$); total motor power capacity (P_w) at different supply voltage U_T. These phases' relationships are shown in Figures 4, 5.

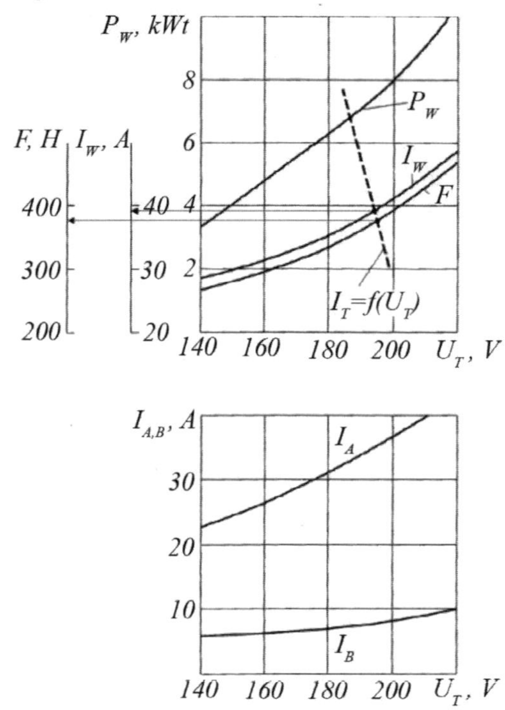

Figure 4. The linear motor's and the feeding transformer's characteristics:
(—) – the motor's characteristics; (----) – the transformer's approximate external inversed characteristic.

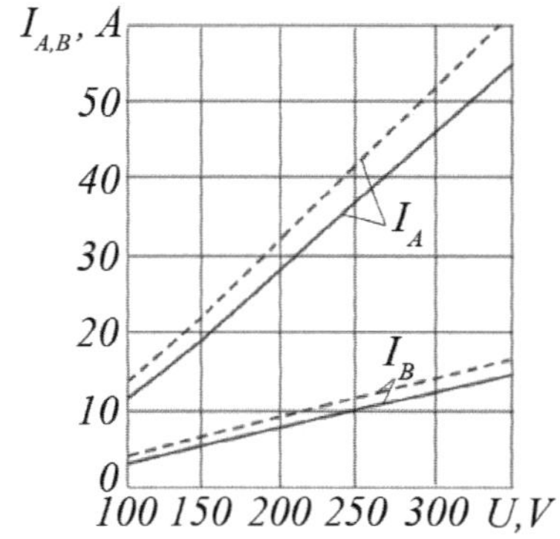

Figure 5. The windings current of the linear motor with different secondary elements:
(—) — a secondary element with a combined layer; (---) — a secondary element with a total cooper layer.

However, these dependencies were applicable when power to the motor was supplied from a stationary source of considerable power, and they gave no answer to the question of what parameters will have the motor when it is powered from a source of limited power - low-power transformer, for example, OM- type transformer. Therefore, graphs of experimental dependencies F, $I_W = f(U_T)$ were subjected to application of the calculated "upside-down" external volt-ampere curve $I_T = f(U_T)$ (see Figure 4) of OM-type transformer, being calculated according to the recommendations of the monograph [7].

The intersection points of this external curve with the curves F, $I_{W,T} = f(U_T)$ allow us to determine the real value of the pulling force and the power consumption of the motor in the 'motor - power supply transformer' system, during a significant short-term transformer overload.

The resulting graph-analytical calculation values of the force and of the motor's current comply with the specifications for high-voltage loading switch.

Practical testing of the system of 'linear motor - transformer power supply with the limited power' at the existing facility proved its good performances and compliance of the calculated and experimental data with an accuracy of about 5%.

Based on the proposed methodology of graph-analytical determination of pulling force and current of linear induction motors we also investigated the possibility of single-phase LIM to be powered from a group of transformers of NOL – type which were connected in parallel to keep the overload mode.

4. Conclusion

The work we did, at the first step clearly shows the ability to power the linear actuators on the basis of LIMs from the high-voltage power lines through individual low-power step-down transformers in the short term overload mode.

In summary, we can say that the graph-analytical motors parameters determination methodology, developed during this work, is useful for practical engineering calculations and it is to be put to a number of uses, in particular of LIMs application in the short-term modes.

In accordance with the foregoing, the author is ready to consider the proposals which are concerned about the development of similar linear electric drives for the different systems which are similar in design for the installations described above.

There is a possibility of trade, on the principle of "as is", of already developed installations with nonconventional electric machines of different movement (for example, with motors for Linear Electric Drives, or for Rotational Electric Drives, or for Disk Drives, or for Electric Rotationally Oscillatory Drives, etc.) with open or close magnetic circuit. A brief description of these motors for Electric Drives of some systems and installations is presented in [1- 6].

For those, who are interested in more detailed information concerning the theory and practice of LIMs, the author can offer English or Russian versions of the lecture's course of 8-16 hours [8], lecturer: Dr. Vasilii TIUNOV, Certified International Engineering Educator (ING-PAED IGIP, Register: RUS-35). The illustrative material of the course is contained in more than 600 slides for presentation, which might be very helpful for engineers and research workers, for organizations of the staff development, postgraduates' and students' study. The course or its fragments were presented and received a high appreciation in Russia, in Austria, in China, in Spain, and in the USA (in four universities of California and at the meeting of Los Angeles Chapter of IEEE Industry Application Society) during the author's lecturing and presentations in the USA.

References

[1] Shulakov N V, Tiunov V V, Iurin A S, Ogarkov E M and Liubimov E V 2005 *Russia. The Department of Electrical Engineering and Electro-Mechanics of the Perm National Research Polytechnic University: the Past and the Present* (City of Perm: The PNRPU publishing house) 185 p

[2] Ogarkov E M 2003 *Russia. The quasi-three-dimensional theory of linear induction motors* (City of Perm: The Perm State Technical University) 240 p

[3] Tiunov V V 2014 *Czech Republic*. Electromagnetic fields, characteristics and practical structures of linear induction machines with short operating body *J. Acta Technica* 59 vol 2 pp 115-134

[4] Rusov V A, Tiunov V V 1976 *Russia.* Method of electromagnetic field calculating for a tubular linear induction motor *Special electric machines and systems of electric drive* (*Collection of Scientific works* vol 194) (City of Perm: Perm Polytechnic Institute) pp 68-78

[5] Tiunov V V 2014 *Russia.* Induction electric machines with open magnetic circuit: linear and other special types of electric motors (*Preprint of the PNRPU Information List*) 2 p

[6] Tiunov V V, Chikarenko S K 2014 *Russia.* Linear electric motors for electromechanical installations: constructions, engineering calculations and linguistic description *Proc. of ASRTU Small Satellites Students' Innovation Forum* (City of Blagoveshensk: the AmSU publishing house) pp. 59-61

[7] Ivanov-Smolenskiy A V 1980 *Russia. Electric machines* (Moscow: 'Energy' publishing house) 592 p

[8] Tiunov V V 2014 *Russia.* Induction electric machines with open magnetic circuit: linear and other special types of electric motors (manuscript and PPt-presentation of the author's lecture course) (City of Perm: the PNRPU) 700 p